High-Energy Physics and Cosmology

Celebrating the Impact of 25 Years of Coral Gables Conferences

High-Energy Physics and Cosmology

Celebrating the Impact of 25 Years of
Coral Gables Conferences

Edited by

Behram N. Kursunoglu

Global Foundation, Inc.
Coral Gables, Florida

Stephan L. Mintz

Florida International University
Miami, Florida

and

Arnold Perlmutter

University of Miami
Coral Gables, Florida

Springer Science+Business Media, LLC

Library of Congress Cataloging-in-Publication Data

High-energy physics and cosmology : celebrating the impact of 25 years
  of Coral Gables conferences / edited by Behram N. Kursunoglu,
  Stephan L. Mintz and Arnold Perlmutter.
       p.   cm.
     "Proceedings of an International Conference on Orbis Scientiate
  1997: Celebration of 25 Coral Gables Conferences and their Impact on
  High-Energy Physics and Cosmology since 1964, held January 23-26,
  1997, in Miami Beach, Florida."--T.p. verso.
     Includes bibliographical references and index.
     ISBN 978-1-4613-7464-0   ISBN 978-1-4615-5397-7 (eBook)
     DOI 10.1007/978-1-4615-5397-7
     1. Particles (Nuclear physics)--Congresses. 2. Neutrinos-
  -Congresses. 3. Cosmology--Congresses. 4. Nuclear astrophysics-
  -Congresses.   I. Kursunoğlu, Behram, 1922-   . II. Mintz, Stephan
  L. III. Perlmutter, Arnold, 1928-   . IV. International Conference
  on Orbis Scientiate 1997: Celebration of 25 Coral Gables Conferences
  and their Impact on High-Energy Physics and Cosmology since 1964
  (1997 : Miami Beach, Florida)
  QC793.H55 1997
  539.7'2--dc21   97-41789
                                                                  CIP

Proceedings of an International Conference on Orbis Scientiate 1997: Celebration of 25 Coral Gables
Conferences and their Impact on High-Energy Physics and Cosmology since 1964,
held January 23 – 26, 1997, in Miami Beach, Florida

This volume was taken from a series of conferences sponsored by Global Foundation, Inc.,
Coral Gables, Florida

ISBN 978-1-4613-7464-0

© 1997 Springer Science+Business Media New York
Originally published by Plenum Press, New York in 1997
Softcover reprint of the hardcover 1st edition 1997

High-Energy Physics and Cosmology:Celebrating the Impact of 25 Years of Coral Gables. . .
  (Kursunoglu, Mintz, Perlmutter, eds.)

PREFACE

The 25th Coral Gables Conference was the culmination of the series that was begun in 1964. The conferences evolved under the titles that include: Symmetry Principles at High Energy; Fundamental Interactions; Orbis Scientiae; and, occasionally, Unified Symmetry in the Small and in the Large. There was a pause after the 20th meeting in 1983 which was dedicated to P.A.M. Dirac. The conferences were resumed in 1993. Some of the reminiscences involved the absence of great minds who attended these meetings in the past and who were no longer with us. The list includes, just to name a few: Julian Schwinger, Robert Oppenheimer, Lars Onsager, Robert Hofstater, Abdus Salam, Richard Feynman, Stanislov Ulam, P.A.M. Dirac, Lord C.P. Snow, Eugene P. Wigner, Vladimir K. Zworykin, and Dixie Lee Ray. Most of these people were among the architects of modern physics and had participated in many of the early Coral Gables Conferences. We miss them.

These conferences have contributed to the progress in high energy physics and cosmology. This year, again, papers were presented on familiar topics, such as neutrino masses, age and total mass of the universe, on the nature of dark matter, and on supersymmetry. The latter has now become a perennial issue. Like the weather, we all talk about it, but, so far cannot do anything to affect it. Another favorite subject was so-called monopoles, which theoretically participate in phenomena like condensation, confinement of electric charge, confinement of monopoles themselves, etc.

Our 1998 meeting will be focused on the subject matter of the physics of mass. This will include, the masses of the neutrinos, of elementary particles, of black holes, and of the universe at large. "What is the origin of mass?" is also among the questions that will be posed at our next meeting.

The Trustees and the Chairman of the Global Foundation, Inc., wish to extend special thanks to Edward Bacinich of Alpha Omega Research Foundation for his continuing generous support including the 1997 Orbis Scientiae.

The Editors
Coral Gables, Florida

ABOUT THE GLOBAL FOUNDATION, INC.

The Global Foundation, Inc., utilizes the world's most important resource... people. The Foundation consists of distinguished men and women of science and learning, and of outstanding achievers and entrepreneurs from industry, governments, and international organizations, along with promising and enthusiastic young people. These people convene to form a unique and distinguished interdisciplinary entity to address global issues requiring global solutions and to work on the frontier problems of science.

Global Foundation Board of Trustees

Global Foundation's Recent Conference Proceedings

Making the Market Right for the Efficient Use of Energy
Edited by: Behram N. Kursunoglu
Nova Science Publishers, Inc., New York, 1992

Unified Symmetry in the Small and in the Large
Edited by: Behram N. Kursunoglu, and Arnold Perlmutter
Nova Science Publishers, Inc., New York, 1993

Unified Symmetry in the Small and in the Large - 1
Edited by: Behram N. Kursunoglu, Stephen Mintz, and Arnold Perlmutter
Plenum Press, 1994

Unified Symmetry in the Small and in the Large - 2
Edited by: Behram N. Kursunoglu, Stephen Mintz, and Arnold Perlmutter
Plenum Press, 1995

Global Energy Demand in Transition: *The New Role of Electricity*
Edited by: Behram N. Kursunoglu, Stephen Mintz, and Arnold Perlmutter
Plenum Press, 1996

Economics and Politics of Energy
Edited by: Behram N. Kursunoglu, Stephen Mintz, and Arnold Perlmutter
Plenum Press, 1996

Neutrino Mass, Dark Matter, Gravitational Waves, Condensation Of Atoms And Monopoles, Light Cone Quantization
Edited by: Behram N. Kursunoglu, Stephen Mintz, and Arnold Perlmutter
Plenum Press, 1996

Technology for the Global Economic, Environmental Survival and Prosperity
Edited by: Behram N. Kursunoglu, Stephen Mintz, and Arnold Perlmutter
Plenum Press, 1997

25th Coral Gables Conference on High Energy Physics and Cosmology
Edited by: Behram N. Kursunoglu, Stephen Mintz, and Arnold Perlmutter
Plenum Press, 1997

Contributing Co-Sponsors of the Global Foundation Conferences, Past and Present:

CONFERENCE PROGRAM

8:00 AM - Noon   REGISTRATION - *Upper Lobby*

1:30 PM   **SESSION I:**   **REMINISCENCES ABOUT THE PAST CORAL GABLES CONFERENCES AND THEIR IMPACT ON THE EVOLUTION OF PHYSICS AND COSMOLOGY**

              Moderator: **BEHRAM N. KURSUNOGLU,** Global Foundation, Inc., Coral Gables, Florida

       Round Table Dissertators: **BEHRAM N. KURSUNOGLU**
              MORTON HAMERMESH, University of Minnesota
              SYDNEY MESHKOV, CALTECH
              ARNOLD PERLMUTTER, University of Miami
              KATSUMI TANAKA, Ohio State University

              Annotators: **ALAN KRISCH,** University of Michigan
              FRED ZACHARIASAN, CALTECH

3:15 PM   **COFFEE BREAK**

3:30 PM   **SESSION II:**   **PHYSICS, ASTROPHYSICS, AND COSMOLOGY OF NEUTRINOS**

              Moderator: **WILLIAM LOUIS,** Los Alamos National Laboratory

              Dissertators: **GEORGE FULLER,** University of California, San Diego
              "Astrophysical Insights Into Neutrino Masses And Mixings"
              GUIDO DREXLIN, Institut fur Kernphysik, Germany
              "Results From The Neutrino Oscillation Search With KARMEN"
              ION STANCU, Los Alamos National Laboratory
              "LSND Neutrino Oscillation Results"
              FRANCOIS VANNUCCI, Lab. De Physique Nucleaire Hautes Energies, Paris
              "Neutrino Oscillations At Accelerators"
              ED KEARNS, Boston University
              "Results From Super Kamiokande"
              AKSEL HALLIN, Queen's University, Canada
              "The Sudbury Neutrino Observatory"

              Annotators: **DARREL SMITH,** Embry-Riddle Aeronautical University, Arizona

       Session Organizer: **WILLIAM LOUIS**

6:30 PM **ORBIS SCIENTIAE ADJOURNS FOR THE DAY**

8:30 AM	SESSION III:	**FROM THE BIRTH OF THE UNIVERSE TO THE *CMBR* ERA**
	Moderator:	**EDWARD KOLB,** FERMI Lab and University of Chicago "Inflation and Its Aftermath"
	Dissertators:	**SCOTT DODELSON,** FERMI Lab "The End of Cosmic Confusion" **LUCA AMENDOLA,** Osservatorio Astronomico di Roma "Redshift Surveys and the Scale of Homogeneity" **JANNA LEVIN,** UC Berkeley "Chaos and the Big Bang"
	Annotators:	**KATHERINE FREESE,** University of Michigan
	Session Organizer:	**EDWARD KOLB**
10:30 AM	**COFFEE BREAK**	
10:45 AM	SESSION IV:	**SEARCH FOR GRAVITATIONAL WAVES**
	Moderator:	**SYDNEY MESHKOV,** CALTECH "Current Status of LIGO"
	Dissertators:	**SAMUEL FINN,** Northwestern University "Observing the Graviton Spin"
	Annotators:	**BEHRAM N. KURSUNOGLU**
	Session Organizer:	**SYDNEY MESHKOV**
12:30 PM	**LUNCH BREAK**	
1:30 PM	SESSION V:	**PROGRESS ON NEW AND OLD IDEAS - I**
	Moderator:	**DON B. LICHTENBERG,** Indiana University "Exotic Hadrons"
	Dissertators:	**PAUL FRAMPTON,** University of North Carolina "Cabibbo Mixing and the Search for CP Violation" **ALAN KOSTELECKY,** Indiana University "CPT Violation and the Standard Model" **PRAN NATH,** Northeastern University, Boston, Massachusetts "Non-Universal Soft SUSY Breaking, SUSY WIMPS And Dark Matter" **MARK SAMUEL,** Oklahoma State University "Going to Higher Order: The Hard Way and the Easy Way - The Agony and the Ecstasy (#7)" **MARCELO R. UBRIACO,** University of Puerto Rico "Anyonic Behavior of Quantum Group Bosonic and Fermionic Systems"

GEOFFREY WEST, LANL
"The Origin of Universal Scaling Laws in Biology and the
Physics of the Cardiovascular and Respiratory Systems"

Annotators: GERALD GURALNIK, Brown University

Session Organizer: DON B. LICHTENBERG

3:30 PM   **COFFEE BREAK**

3:45 PM   **SESSION V:**   **PROGRESS ON NEW AND OLD IDEAS - I**
             (continued)

5:30 PM   Orbis Scientiae Adjourns for the Day

SATURDAY, January 25, 1997

8:30 AM   **SESSION VI:**   **ROUND-TRIP BETWEEN COSMOLOGY AND
                                   ELEMENTARY PARTICLES**

             Moderator: **KATHERINE FREESE**
                             "Low Mass Stars And The Baryonic Dark Matter Of The
                             Universe"

             Dissertators: **DAVID CLINE**, UCLA
                             "How the Search for SUSY Dark Matter is Going - Prospects for
                             Discovery"
                             JIM REIDY, University of Mississippi
                             "CP Violation and The BeBar Detector Program at SLAC"
                             VIGDOR TEPLITZ, Southern Methodist University
                             "Use of Seismology to Search for Strange Quark Matter"

             Annotators: **L.G. RATNER**, University of Michigan

             Session Organizer: **KATHERINE FREESE**

10:30 AM   **COFFEE BREAK**

10:45 AM   **SESSION VII:**   **GUAGE SYMMETRIES, GRAVITY AND STRINGS**

             Moderator: **LOUISE DOLAN**, University of North Carolina
                             "Gauge Symmetry and Discrete Liouville States"

             Dissertators: **MURAT GÜNAYDIN**, Penn State University, Pennsylvania
                             "Conformal Symmetry and Duality Groups of Superstring and
                             Supergravity Theories"
                             FREYDOON MANSOURI, University of Cincinnati
                             "Exact Local Symmetry, Absence of Superpartners, and Non-
                             Commutative Geometries"

S. RAJEEV, University of Rochester, New York
"Four Dimensional Scalar Conformal Field Theory"

Annotators: **PRAN NATH**

Session Organizer: **LOUISE DOLAN**

12:15 PM   **LUNCH BREAK**

1:30 PM   **SESSION VIII:**   **LIGHT CONE QUANTIZATION**

Moderator: **STEPHEN PINSKY,** Ohio State University
"Light-Cone Approach To Large N Matrix Models"

Dissertators: **JOHN HILLER,** University of Minnesota
"Non-Perturbative Renormalization In Light-Cone Quantization"
MARTINA BRISUDOVA, Ohio State University
"QCD Bound States in a Light-Front Hamiltonian Approach"

Annotators:

3:00 PM   **COFFEE BREAK**

3:15 PM   **SESSION VIII:**   **LIGHT CONE QUANTIZATION**
                (continued)   **Dissertator Presentations Continue:**

Moderator: **ROBERT PERRY,** Ohio State University

Dissertators: **BILLY D. JONES,** Ohio State University
"The QED Bound State Problem in a Light Front Hamiltonian Approach"
ROBERT PERRY
"Constituent Bound States In Light-Front QCD"
KATSUMI TANAKA AND ABE AUBRECHT, Ohio State University
"Mesons in 2D QCD on the Light Cone"

Annotators: **STEPHEN PINSKY**

Session Organizer: **STEPHEN PINSKY**

5:30 PM   Orbis Scientiae Adjourns for the Day

7:30 PM   Conference Banquet (Courtesy of Maria and Edward Bacinich) - *Mona Lisa Ballroom*

8:30 AM   **SESSION IX:**   **PROGRESS ON NEW AND OLD IDEAS - II**

Moderator: **FRED ZACHARIASEN,** CALTECH

	Dissertators:	**RICHARD ARNOWITT,** Texas A & M University "Seeing Planck Scale Physics at Accelerators" **GERALD CLEAVER,** Ohio State University "Results of an SO(10) String GUT Search" **STEPHEN MINTZ,** Florida International University "Neutrino Reactions in Nuclei" **PIERRE RAMOND,** University of Florida "The Anomalous U(1) and Low Energy" **INA SARCEVIC,** University of Arizona "Detection Of Neutrinos From Topological Defects Formed In The Early Universe"
	Annotators:	**ZACHARY GURALNIK,** Princeton University
	Session Organizer:	
10:30 AM	**COFFEE BREAK**	
11:00 AM	**SESSION X:**	**CURRENT EXPERIMENTS IN HIGH ENERGY PHYSICS**
	Moderator:	**JOSEPH LANNUTTI,** Florida State University
	Dissertators:	**WOLFGANG ADAM,** LEP-DELPHI Collaboration at CERN "Searches for New Particles During the High Energy Run of LEP" **HYWEL PHILLIPS,** LEP-DELPHI Collaboration at CERN "W Physics at LEP 2" **SUDHINDRA MANI,** University of California at Davis "Precision Tests of the Standard Model in D0"
	Annotators:	Participants
	Session Organizer:	**TONI BARONCELLI,** CERN
12:30 PM	**SESSION X:** **(continued)**	**GENERAL DISCUSSION OF SESSION X** **a. Theoretical Comments** **b. Proposed Future Experiments**

1:30 PM   ORBIS SCIENTIAE 1997 ADJOURNS

CONTENTS

SECTION IV
Round Trip Between Cosmology and Elementary Particles

SECTION V
Gauge Symmetries, Gravity and Strings

SECTION VI
Light Cone Quantization

SECTION VII
Current Experiments in High Energy Physics

SECTION I
LAUNCHING OF THE CORAL GABLES CONFERENCES ON HIGH ENERGY PHYSICS AND COSMOLOGY

25TH CORAL GABLES CONFERENCE PRESENTATION

Thursday, January 23, 1997

Behram N. Kursunoglu

THE LAUNCHING OF THE CORAL GABLES CONFERENCES ON HIGH ENERGY PHYSICS AND COSMOLOGY AND THE ESTABLISHMENT OF THE CENTER FOR THEORETICAL STUDIES AT THE UNIVERSITY OF MIAMI*

Behram N. Kursunoglu

Global Foundation, Inc., Coral Gables, Florida
(kursungf@netrunner.net)

On this occasion of the 25th Coral Gables Conference I shall present the history of initial formative phase of these conferences. Because of the important role played in launching the conferences and in establishing the Center for Theoretical Studies (CTS), the narration about J. Robert Oppenheimer will be included in some detail. The first Coral Gables Conference on "Symmetry Principles at High Energy" in January 1964 was followed by the second Coral Gables Conference (CGC.II) in January 1965, "Symmetry Principles at High Energy", was held with even greater enthusiasm and dedication than

From left to right: Sevda Kursunoglu, Dr. J. Robert Oppenheimer, and Dr. Abdus Salam.

CGC.I for which, as will be seen in the following presentation, there were quite a few reasons. First I would like to quote a remark of P.A.M. Dirac's (a Nobel Laureate in physics) when he asked Oppenheimer, "How do you write poetry and at the same time work on theoretical physics?" I do not remember Oppenheimer's answer as was related to us by Dirac. However, I would like to quote from Dirac's Oppenheimer Memorial Prize acceptance speech given in March 1969 during the awards ceremonies at the Center for Theoretical Studies:

> "I was a great friend and admirer of Oppenheimer. I knew him for more than forty years. There was a time when we were young students together at Göttingen. We both stayed in the same pension. We both had the same interests, going to the same lectures and we found that we also had interests outside the lectures. We both liked

---

* This paper is, with some modification, taken from the typescript of the author's book with the provisional title "The Ascent of Gravity: Einstein's Triumph, Paradigms, and Contemporary Physicists" which is to be published in 1997 or early 1998.

to go for long walks and occasionally took a day long walk across country together. Since these early days, I have met Oppenheimer on many occasions and I have been able to see what admirable qualities he had, particularly as a chairman for a discussion or colloquium. He had a very quick mind which enabled him to pick on the main point of the issue and if there was something which the lecturer couldn't explain very well, or some member of the audience was asking a question which he could not formulate very clearly, Oppenheimer would frequently jump into the breach and explain in lucid language just what was needed in order to bring the point clearly home to everybody and enable the discussion to proceed on clarified lines. It is a great loss to science and to us all that he died so young and I especially feel his loss because of the great personal friendship I had for him."

If a physicist can make one basic contribution to physics then he or she belongs to the category of outstanding physicists. Oppenheimer's two most important contributions to physics may be summarized as follows: (1) by accepting Herman Weyl's interpretation of Dirac's relativistic theory of the electron which predicted two particles with equal mass, he showed that one of these particles ought to carry, opposite to that of an electron, a positive electric charge and that the new particle was not a proton; (2) in 1938 he deduced from general relativity the concept of gravitational collapse. It must also be emphasized that Oppenheimer was one of the most effective leaders in bringing theoretical physics to the United States. He trained many outstanding students who became leaders in their own fields and quite a few of them were awarded Nobel Prizes.

There is another side to Oppenheimer. He was a great leader among physicists in the United States. He was in charge of the physics and technology of the Manhattan project where the first atomic bomb was designed and produced in 1945 at the Los Alamos Scientific Laboratory. He was involved in a mega project of the US Government that changed the world forever. He was the envy of many people. He had to carry the burden of having presided over something that shook the world. From there on, anything that was said or done by Oppenheimer was subject to unlimited scrutiny by his "friends," by the administration, the Congress, and above all, the media. He was made controversial: was he a leftist? Did he give any atomic secrets to the Soviet agents? Was Oppenheimer a communist? Oppenheimer's security clearance was revoked and he resigned from the US Atomic Energy Commission's Advisory Board, and was appointed by the trustees as the director of the prestigious Institute for Advanced Study at Princeton, New Jersey. This was, as briefly described above, his *curriculum vitae* who at one time was one of the most distinguished men in the world. When I visited him at the Institute he had accepted two invitations from me: (1) to attend and present a paper in the CGC.II held in January 1965, (2) to accept membership on the Scientific Council of the soon to be established Center for Theoretical Studies at the University of Miami. He asked me to stay on for lunch in the Institute cafeteria and meet some of the trustees. I accepted with great pleasure. I always addressed him as Professor Oppenheimer which he objected to by saying that I should use a better alternative. However, he never let me know what was that "better alternative." He now was, for the first time, visiting the University of Miami at Coral Gables.

Oppenheimer was welcomed to the University of Miami Campus very much like a head of state, not like a modest theoretical physicist. While walking on the campus toward the Lowe Art Museum, the conference site, he was followed by a crowd of a few hundred people that included, besides conference participants, members of the University of Miami faculty and administration led by Henry King Stanford, the new president of the University of Miami. Stanford, who had spent the year 1957 in Ankara, Turkey in the New York University Educational Mission, was able to speak Turkish, and had become a Turcophile. In the Lowe Art Museum Oppenheimer presided over the first session, and made a presentation on the "Symmetry Principles." The participants included, besides those who attended CGC.I, Gregory Breit, B. D'Espagnat, W.R. Gruner (NSF), B.W. Lee, D.R. Lehman (USAFOSR), H.J. Lipkin, L. Michel, Yoichiro Nambu, Peter Rosen, M.A.

Ruderman, Abdus Salam, E.C.G. Sudarshan, Alvin M. Weinberg (after dinner speaker), G.Zweig, Bruno Zumino, J. Nuyts, W.D. McGlinn, M. Mayer, A.J. MacFarlane, J.S. Lomont, R. Hermann, E. Guth, G. Goldhaber, H. Ekstein, and F. Coester.

Following Oppenheimer's presentation, Abdus Salam was the second speaker proposing a new theory of strong interaction symmetry based again on a new way of unifying space time and internal symmetries. This time Salam proposed another symmetry group symbolized by $\tilde{U}(12)$. Unfortunately it turned out to be another good example of Hodja's "yogurt project". Nasrettin Hodja was a Turkish folklore hero who lived in 13th century middle Anatolia. One morning a woodcutter saw Hodja by the edge of a lake, throwing quantities of yeast into the water. "What the devil are you doing, Hodja?" he asked. Hodja looked up sheepishly and replied, "I am trying to make all the lake into yogurt." The woodcutter laughed and said. "Fool, such a plan will never succeed." Hodja remained silent for a while, and stroked his beard. Then he replied, "But just imagine if it should work!"

However, I was very pleased to see that the interest in the idea of unifying internal and space-time symmetries I proposed in CGC.I was continuing. When Salam arrived in Coral Gables his first sentence to me was to ask if Oppenheimer was coming to the conference. He then continued modestly(!) by saying that "Oppenheimer must be grateful to Kursunoglu for bringing the solution of the symmetry problem to the USA," i.e., Salam, being invited by me, brought the solution with him in his $\tilde{U}(12)$. Other presentations on the same subject of symmetries unification were made by Sudarshan, Lipkin, Ne'eman, Bardakci, Freund, Nambu, Michel, Zachariasen, and Zweig.

The 1965 conference banquet was held in the plush Kings Bay Yacht Club located about five miles southeast of the University. Despite my disinclination to have an after dinner speaker, I invited Alvin M. Weinberg, Director of Oak Ridge National Laboratory to address our conference. Weinberg had nothing to do with high energy physics. As a result of this shortcoming, his presentation seemed like an outsider's intrusion into the gathering of the high energy physicists. Oppenheimer compared it to a managerial presentation and gave a quotation from Jefferson which I do not remember.

I had invited many faculty members and the president of the University, Henry King Stanford to the conference dinner. I requested some of them to pass by the University Inn and give a ride to some of the conference participants. Murray and Rose Mantell from the School of Engineering gave a ride to Oppenheimer and H. Primakoff. They often reminisced the occasion with pride and fondness. For the 1966 CGC.III, Oppenheimer was the after dinner speaker in, again, Kings Bay Yacht Club. The topic of his talk was "Einstein." The speech was published in the proceedings of CGC.III by W.H. Freeman and Company, San Francisco and London. Editors B.N. Kursunoglu et al. 1966. The speech is about Einstein, but it is also, in part, about Oppenheimer himself since it throws some light on the beliefs not only of Einstein's but that of Oppenheimer's too. Therefore, I am including it in its entirety.

Talk Delivered at the Banquet of the Conference
January 19, 1966
Kings Bay Yacht Club, Miami

by

J. Robert Oppenheimer

UNESCO decided to celebrate the tenth anniversary of Einstein's death, the fiftieth of his discovery of general relativity, last month in Paris. They asked me to come and to speak briefly.

There were many other speakers: de Broglie, Heisenberg, Gonseth, who had been a student in Switzerland in 1905, and told wonderfully of what that was like.

The New York Times published a brief account of my talk, and since then a number of colleagues have suggested that I had been out of mind. I thought you might like to hear what I said. It is mercifully brief. Here it is.

Though I knew Einstein for two or three decades, it was only in the last decade of his life that we were close colleagues and something of friends. But I thought that it might be useful, because I am sure that it is not too soon--and for our generation perhaps almost too late--to start to dispel the clouds of myth and to see the great mountain peak that these clouds hide. As always, the myth has its charms; but the truth is far more beautiful.

Late in his life, in connection with his despair over weapons and wars, Einstein said that if he had to live it over again he would be a plumber. This was a balance of seriousness and jest that no one should now attempt to disturb. Believe me, he had no idea of what it was to be a plumber; least of all in the United States, where we have a joke that the typical behavior of this specialist is that he never brings his tools to his crises; Einstein was a physicist, a natural philosopher, the greatest of our time.

What we have heard, what you all know, what is the true part of the myth is his extraordinary originality. The discovery of quanta would surely have come one way or another, but he discovered them. Deep understanding of what it means that no signal could travel faster than light would surely have come; the formal equations were already known; but this simple, brilliant understanding of the physics could well have been slow in coming, and blurred, had he not done it for us. The general theory of relativity which, even today, is not well proved experimentally, no one but he would have done for a long, long time. It is in fact only in the last decade, the last years, that one has seen how a pedestrian and hard working physicist, or many of them, might reach that theory and understand this singular union of geometry and gravitation, and we can do even that today only because some of the a priori open possibilities are limited by the confirmation of Einstein's discovery that light would be deflected by gravity.

Yet there is another side besides the originality. Einstein brought to the work of originality deep elements of tradition. It is only possible to discover in part how he came by it, by following his reading, his friendships, the meager record that we have. But of these deep seated elements of tradition--I will not try to enumerate them all; I do not know them all--at least three were indispensable and stayed with him.

The first is from rather beautiful but recondite part of physics that is the explanation of the laws of thermodynamics in terms of the mechanics of large numbers of particles, statistical mechanics. This was with Einstein all the time. It was what enabled him from Planck's discovery of the law of black body radiation to conclude that light was not only waves but particles, particles with an energy proportional to their frequency and momentum determined by their wave-number, the famous relations that de Broglie was to extend to all matter, to electrons first and then clearly to all matter.

It was this statistical tradition that led Einstein to the laws governing the emission and absorption of light by atomic systems. It was this that enabled him to see the connection between de Broglie's waves and the statistics of light-quanta proposed by Bose. It was this that kept him an active proponent and discoverer of the new phenomena of quantum physics up to 1925.

The second and equally deep strand--and here I think we do know where it came from--was his total love of the idea of a field: the following of physical phenomena in minute and infinitely subdividable detail in space and in time. This gave him his first great drama of trying to see how Maxwell's equations could be true. They were the first field equations of physics; they are still true

8

today with only very minor and well-understood modifications. It is the tradition which made him know that there had to be a field theory of gravitation, long before the clues of that theory were securely in his hand.

The third tradition was less one of physics than of philosophy. It is a form of the principle of sufficient reason. It was Einstein who asked what do we mean, what can we measure, what elements in physics are conventional? He insisted that those elements that were conventional could have no part in the real predictions of physics. This also had roots: for one the mathematical invention of Riemann, who saw how very limited the geometry of the Greeks had been, how unreasonably limited. But in a more important sense, it followed from the long tradition of European philosophy, you may say starting with Descartes--if you wish you can start it in the Thirteenth Century, because in fact it did start then--and leading through the British empiricists, and very clearly formulated, though probably without influence in Europe, by Charles Pierce: one had to ask how do we do it, what do we mean, is this just something that we use to help ourselves in calculating, or is it something that we can actually study in nature by physical means. For the point here is that the laws of nature not only describe the results of observations, but the laws of nature delimit the scope of observations. That was the point of Einstein's understanding of the limiting character of the velocity of light; it also was the nature of the resolution in quantum theory, where the quantum of action, Planck's constant, was recognized as limiting the fineness of the transaction between the system studied and the machinery used to study it, limiting this fineness in a form of atomicity quite different from and quite more radical than any that the Greeks had imagined or that was familiar for the atomic theory of chemistry.

In the last years of Einstein's life, the last twenty-five years, his tradition in a certain sense failed him. They were the years he spent at Princeton and this, though a source of sorrow, should not be concealed. He had a right to that failure. He spent those years first in trying to prove that the quantum theory had inconsistencies in it. No one could have been more ingenious in thinking up unexpected and clever examples; but it turned out that the inconsistencies were not there; and often their resolution could be found in earlier work of Einstein himself. When that did not work, after repeated efforts, Einstein had simply to say that he did not like the theory. He did not like the elements of indeterminacy. He did not like the abandonment of continuity or of causality. These were things that he had grown up with, saved by him, and enormously enlarged; and to see them lost, even though he had put the dagger in the hand of the assassin by his own work, was very hard on him. He fought with Bohr in a noble and furious way, and he fought with the theory which he had fathered but which he hated. That is not the first time it has happened in science.

He also worked with a very ambitious program, to combine the understanding of electricity and gravitation in such a way as to explain what he regarded as the semblance--the illusion--of discreteness, of particles in nature. I think that it was clear then, and believe it to be obviously clear today, that the things that this theory worked with were too meager, left out too much that was known to physicists but had not been known much in Einstein's student days. Thus it looked like a hopelessly limited and historically rather accidentally conditioned approach. Although Einstein commanded the affection, or, more rightly, the love of everyone for his determination to see through his program, he lost most contact with the profession of physics, because there were things that had been learned which came too late in life for him to concern himself with them.

Einstein was indeed one of the friendliest of men. I had the impression that he was also, in an important sense, alone. Many very great men are lonely; yet I had the impression that although he was a deep loyal friend, the stronger human affections played a not very deep or very central part in his life taken as a whole. He had of course many disciples, in the sense of people who, reading his work or hearing it taught by him, learned from him and had a new view of physics, of the philosophy of physics, of the nature of the world that we live in. But he did not have, in the technical jargon, a school. He did not have very many students who were his concern as apprentices and disciples. And there was an element of the lone worker in him, in sharp contrast to the teams we see today, and in sharp contrast to the highly cooperative way in which some other parts of science have developed. In later years, he had people working with him. They were typically called assistants and they had a wonderful life. Just being with him was wonderful. His secretary had a wonderful life. The sense of grandeur never left him for a minute, nor his sense of humor. The assistants did one thing which he lacked in his young days. His early papers are paralyzingly beautiful, but there are many errata. Later there were none. I had the impression that, along with its miseries, his fame gave him some pleasures, not only human pleasure of meeting people but the extreme pleasure of music played not only with Elizabeth of Belgium but more with Adolphe Busch, for he was not that good a violinist. He loved the sea and he loved sailing and was always grateful for a ship. I remember walking home with him on his seventy-first birthday. He said, "You know, when it's once been given to a man to do something sensible, afterward life is a little strange."

Einstein is also, and I think rightly, known as a man of very great goodwill and humanity. Indeed, if I had to think of a single word for his attitude towards human problems, I would pick the Sanscrit word "Ahinsa," not to hurt, harmlessness. He had a deep distrust of power; he did not have that convenient and natural converse with statesmen and men of power that was quite appropriate to Rutherford and to Bohr, perhaps the two physicists of this century who most nearly rivalled him in eminence. In 1915, as he made the general theory of relativity, Europe was tearing itself to pieces and half losing its past. He was always a pacifist. Only as the Nazis came into power in Germany did he have some doubts, as his famous and rather deep exchange of letters with Freud showed, and began to understand with melancholy and without true acceptance that, in addition to understanding, man sometimes has a duty to act.

After what you have heard, I need not say how luminous was his intelligence. He was almost wholly without sophistication and wholly without worldliness. I think that in England people would have said that he did not have much "background," and in America that he lacked "education." This may throw some light on how these words are used. I think that this simplicity, this lack of clutter and this lack of cant, had a lot to do with his preservation throughout of a certain pure, rather Spinoza-like, philosophical monism, which of course is hard to maintain if you have been "educated" and have a "background." There was always with him a wonderful purity at once childlike and profoundly stubborn.

Einstein is often blamed or praised or credited with these miserable bombs. It is not in my opinion true. The special theory of relativity might not have been beautiful without Einstein; but it would have been a tool for physicists, and by 1932 the experimental evidence for the inter-convertibility of matter and energy which he had predicted was overwhelming. The feasibility of doing anything with this in such a massive way was not clear until seven years later, and then almost by accident. This was not what Einstein really was after. His part was that in creating an intellectual revolution, and discovering more than any scientist of our time how profound were the errors made by men before then. He did write a letter to Roosevelt about atomic energy. I think this was in part his agony at the evil of the Nazis in part not wanting to harm any one in any way; but I ought to report that that letter had very little effect, and that Einstein himself is really not answerable for all that came later. I believe he so understood it himself.

His was a voice raised with a very great weight against violence and cruelty wherever he saw them and, after the war, he spoke with deep emotion and I believe with great weight about the supreme violence of these atomic weapons. He said at once with great simplicity: no we must make a world government. It was very forthright, it was very abrupt, it was no doubt "uneducated," no doubt without "background;" still all of us in some thoughtful measure must recognize that he was right.

Without power, without calculation, with none of the profoundly political humor that characterized Gandhi, he nevertheless did move the political world. In almost the last act of his life, he joined with Lord Russell in suggesting that men of science get together and see if they could understand one another and avert the disaster which he foresaw for the arms race. The so-called Pugwash movement, which has a longer name now, was the direct result of this appeal. I know it to be true that it had an essential part to play in the Treaty of Moscow, the limited test-ban treaty, *which is a tentative, but to me very precious, declaration that reason might still prevail.*

In his last years, as I knew him, Einstein was Twentieth Century Ecclesiastes, saying with unrelenting and indomitable cheerfulness, "Vanity of vanities, all is vanity."

--End of Speech--

This was an impressive oration. How does Oppenheimer's rhetoric impact on me for the second time 30 years after it was made? His skill and eloquence in public speaking, and his command of the English language were known to be quite unique. It was as if he was speaking to his rivals and benefiting, both by praising and by blasting, one of the greatest, as well as the most famous, physicist in the entire history of science and learning. He also revealed some of the conceit and, perhaps, some of the arrogance which were not unknown to most people. He spoke also like the Boss of Albert Einstein. He spoke of Einstein's loneliness--shared by most great people--which was quite compatible with the discovery of general relativity. Contrary to Oppenheimer's team work approach, general relativity, and quantum theory were not discovered by a team of physicists. Oppenheimer's criticism of Einstein's program to combine electricity and gravity as ambitious was mostly due to his mindset. He attributed Einstein's disadvantages to his "hopelessly limited and historically rather accidentally conditioned approach." This was, he might have thought, a crowd pleasing, and not so unprejudiced utterance. Why the attempt to unify a force like gravity, which interacts with everything, with another long-range force like electricity is more ambitious than the unification or rather the "federation" of the weak and electromagnetic interactions? Rather than the word ambitious, we should say that the unification of gravity and electricity is more fundamental, even if, because of their long-range effects, they are the oldest known interactions. Oppenheimer says that "Einstein left out too much that was known to physicists but had not been known in Einstein's student days." In 1910 Einstein did not know about the relevance of Riemannian geometry to physics but by 1915 he merged geometry and gravity to create the general relativistic theory of gravitation. It is not only irrelevant but what the other physicists know about theoretical physics may have been an inhibition to Einstein's program.

Einstein could have benefited from his rich and rewarding experiences in his kind of physics. Max Von Laue's letter of 1915 to Einstein is a good example which he related to me in November 1953 while I was visiting with him in his home. In the letter, Von Laue warns Einstein by writing that "You are wasting your time with your efforts to explain gravity in terms of geometry. You will not succeed. Even if you do no one will believe you." As I heard it from Einstein himself when he told me that Max Von Laue wrote, two years later, a set of two volumes on Einstein's general relativity. Oppenheimer, in retrospect, spoke quite unfairly about Einstein when he said "The last twenty five years in a certain sense failed him. They were the years he spent at Princeton and this, though a source of sorrow, should not be concealed." I am sure most physicists are convinced that Einstein did not have any unproductive phase in his entire life as a physicist. Einstein's attempts to find inconsistencies in quantum theory, his many dialogues with Niels Bohr have greatly contributed to the proper understanding and interpretation of the quantum theory, even though he objected rather strongly to its indeterministic base. With regard to Einstein's historic letter of 1939 to President Roosevelt, about the atomic bomb, engineered by three Hungarian physicists Leo Szillard, Eugene P. Wigner, and Edward Teller, Oppenheimer had said that the letter had little effect. What did he mean by "little effect?" How could he have known that Einstein's letter to Roosevelt did not have much effect? I believe Einstein letter had made a profound impact in the eventual initiation by the U.S. government of the Manhattan Project.

Oppenheimer's presence along with other distinguished participants, taking into account the presentation of good papers, made the CGC.II a success. Because of the well established good credentials of CGC.I, the CGC.II was supported by the US Atomic Energy Commission, the US

Air Force Office of Scientific Research, the National Aeronautic and Space Administration, and the National Science Foundation. The faculty and the administration of the University of Miami were favorably impressed with all that CGC.II brought to the campus. It was time to look into the future and, therefore, to the continuation of making Coral Gables a place for the gathering of the world class scholars not only in physics but in other sciences also. This thought called for the establishment of a prominent Center open and working all year round. This would also make it possible to be consistent with what I said to Freeman Dyson in New York during the American Physical Society Meeting in January 1962, that Coral Gables will become a world center for the gathering of the great men of science, at least on an annual basis. I began thinking on the most important starting point of finding a proper and inspiring name for the new organization to imply coverage of most theoretical research fields.

Thanks to Oppenheimer's acceptance of membership on the Scientific Council of the new Center to be established, hopefully, before the end of the 1964-65 academic year, it was quite easy to persuade others to join in. Julian Schwinger, already a friend, agreed to serve. Others were Elliott W. Montroll, Robert Marshak, Wendell M. Stanley, and Yoichiro Nambu. I had already chosen to name the new organization the *Center for Theoretical Studies* (CTS) and had held a one hour afternoon meeting, during the conference, of the first *Scientific Council* at the University Inn, located across from the University of Miami. The recommendations included Oppenheimer's suggestion that we should not focus on the subject of theoretical biology since nothing has ever been explained based on it. Otherwise CTS research could include almost everything to which theoretical physics can be applied. After everyone left town, Arnold Perlmutter and I prepared a draft proposal for the establishment of CTS and submitted it to the president and various academic committees, and especially the Faculty Senate of the University of Miami for their approvals. The proposal contained three important promises: (1) CTS, except providing the needed space, would not need financial support from the University, (2) CTS would bring to the University, all year round, world class scientists and provided a list of prospective visiting scientists including: Julian Schwinger*, Willis Lamb*, John Eccles*, Robert Oppenheimer, Robert Mulliken*, Lars Onsager*, Murray Gell-Man*, Robert Hofstater*, Abdus Salam*, Richard Feynman*, Sheldon Glashow*, Steve Weinberg*, Stanislav Ulam, Edward Teller, Hans Bethe*, P.A.M. Dirac*, Lord C.P. Snow, Manfred Eigen*, Eugene P. Wigner*, John Bardeen*, Nikolai Basov*, H.S.W. Massey, Abraham Pais, Henry Primakoff, Norman Ramsey*, Tullio Regge, Vladimir K. Zworykin, Francis Crick*, Donald Glaser*, Wendell Stanley*, C.H. Wadington, Tjalling C. Kopmans*, Dixie Lee Ray, Hans Bremermann, Enrico Clemento, Glenn T. Seaborg*. These twenty-three Nobel laureates, which reached, later on, the level of 35, and other distinguished scientists covered the fields of theoretical physics, chemistry, biology, mathematics, economics, artificial intelligence, arms control, and literature. More than thirty percent of the above listed scientists were not Nobel laureates at the time that the list was made. And (3) CTS would seek and raise research funds in the field of theory only and would encourage joint efforts with relevant departments.

After obtaining the approvals of the Departments of Physics, Chemistry, Biology, Mathematics, Philosophy, History, the School of Engineering, the School of Marine and Atmospheric Sciences, and finally obtaining the endorsement of the University's Academic Planning Committee, the Center's Charter was submitted to the Faculty Senate. The CTS Charter was approved by a unanimous vote of the Faculty Senate on May 17, 1965, and by the Board of Trustees Executive Committee on May 28, 1965. I was appointed by the president of the University as the director of CTS, and in accordance with the charter, did not have to report to any academic unit but only to the president of the University through the Vice President for Academic Affairs. Any other restrictions in my teaching, research, fund raising, and administrative duties, were not part of the campus politics in which, unless provoked by anyone, I tried not to get involved. Now it was time to get the wheels rolling by attracting the above listed people to the

---

* Nobel Laureate

University. The most effective way was to invite one or two distinguished scientists to attract others to be part of the program: (1) to lecture to students and faculty, (2) to participate in a conference organized by CTS, (3) to stay on long enough to write a paper bearing the CTS name in its lists of acknowledgments.

All the scientists listed above came and stayed at CTS. Part of the third floor of the new computer science building was assigned to CTS. There were seven offices to accommodate ten postdoctoral fellows, and visiting professors. The CTS space also included a conference room, a library, the director's office, a secretarial area for four secretaries, an office for an administrative assistant, a supply room, and an office for the Secretary of CTS, Arnold Perlmutter, whose main job was, of course, to serve as professor of physics in the physics department. At a Christmas party in December of 1965 I attempted to introduce a gentleman to Henry King Stanford as the first chemist visitor to CTS. Unfortunately, just at that moment I forgot the chemist's name, then I turned to introduce instead Stanford to the chemist. Well, I forgot his name too. It was an embarrassing experience. From then on whenever I have had to introduce people first I rehearsed their names in my mind well in advance and then do the usual ceremonies. In other circumstances I just let people introduce each other. This transient memory gap reminded me of a similar incident experienced by Eugene Wigner in the summer of 1964 when we were both attending a small meeting organized by Asim Barut in Boulder, Colorado. We were at Barut's party when Wigner came to me with his usual polite disposition and asked me--pointing to our host Barut, whom he has known many years--"Excuse me, please. Could you tell me the name of this gentleman?" When I obliged him he said, "I am so sorry I should not have forgotten his name." Thinking of this I found some consolation for myself. The names of the people are, of course, the most pleasant sounding word for them and anyone who can remember names should make a successful politician, and a good university president. The best example, in my experience, of a man who can remember the name of somebody's second cousin, even four years later, is the president of the University of Miami (from 1962-1981), Henry King Stanford. During his 19 year presidency, the friends of the University had increased ten-fold. Which also resulted in ten-fold increase in its endowment and program support.

The basic *modus operandi* of CTS was based on visiting distinguished scholars and men and women of learning for periods ranging from a week to many months. The approach was contrary to the traditional institutes and centers where permanent appointments are the basis of the institute with some visiting membership to work in the areas of interest to the institute. There were, besides the director, only two long-term appointments at CTS. The first was P.A.M. Dirac who retired from Cambridge University in 1968 and came to CTS in December of 1968 for a few months stay. The other distinguished scientist was Lars Onsager who was awarded the Nobel prize in chemistry in 1970 and retired from Yale University's chemistry department in 1970. He joined CTS in 1971 on a permanent basis as a distinguished university professor. Most of the membership of CTS were visiting distinguished scholars, professors, and postdoctoral fellows. CTS benefited greatly from the mutual attractions of well-known physicists, biologists, chemists, and men and women of letters. The afternoon tea was served in the conference room for visiting members and interested University of Miami faculty everyday. There were 16 comfortable chairs around a large rectangular table. There was also a fairly large blackboard. Conversation ranged from discussing the evolution of the genetic code to the expanding universe, environment, nuclear weapons and arms control, Fermat's theorem in mathematics, how elementary are the elementary particles?, the theory of chaos, ... These afternoon tea meetings were interesting, amusing, and inspiring for all participants.

It was in 1972 when Francis Crick arrived at CTS for a week's stay. When I was a graduate student in Cambridge, Crick and Watson were divining the DNA molecule in the Cavendish Laboratory adjacent to our building, the Arts School, where the physics graduate students' offices

The foursome Nobel laureates:

Professors Francis Crick, Lars Onsager, P. A. M. Dirac, and Willis Lamb, Jan. 16, 1973, are being watched over by Albert Einstein

and that of Dirac's were located. As physicists, we were quite unaware of the research going on in the world famous Cavendish Laboratory. After all, the physicists were, in this atomic era, the smartest people and the masters of the world! We heard nothing about molecular biology and had made no special effort to inquire about it. It was not until 1961 when Crick and Watson were awarded the Nobel Prize for their work on the helical structure of the DNA molecule that we heard of them. We were profoundly impressed with these two past neighbors in the Cavendish Laboratory. When 20 years later Crick entered my office at CTS at 8:30 in the morning I realized how I missed out as a student by not attending the biology colloquia in Cavendish Laboratory. As soon as Crick sat down he began telling me about his latest ideas and especially about the probability of life on earth originating from space! Crick kept on telling me about his ideas. I was unable to find an opportunity to ask a question or say anything. At one point he coughed and that was my opportunity to talk. I said, "You have been telling me about your good ideas without giving me a chance to ask some questions. You came in at 8:30 a.m. and now my watch shows 10:15 a.m. You know, talking and thinking are sometimes orthogonal, I mean when you do only one of them all the time, it comes out at the cost of the other! He replied jokingly, "I do not like to listen all the time." "But," I said, "I do not wish to listen all the time either." I then invited him to join others in the conference room. "There you are. You must tell about your ideas to Schwinger, Dirac. Onsager, Teller, Lamb, and Eccles." I told them I would be back in an hour and if they wished, we could go to the cafeteria for lunch. When I returned, I saw Dirac sleeping, and Teller looking quite depressed, presumably because he had not gotten his turn to speak after Crick. Crick was still talking but the rest looked ready to go to lunch. Crick had a charming and colorful personality and we were greatly impressed with him and his ever present creative disposition. We were also very pleased to have met the charming Mrs. Odile Crick. Sevda and I entertained the couple and others to dinner in our home.

Tem-pus fu-git, yes time does indeed fly. It flies for everyone, everywhere, all the time! The older one ages the more one reevaluates the past time and counts one's regrets or one's triumphs. The greatest triumphs result from the implementation of good ideas affecting and benefiting many people. A good idea leads to change which for the scientist means coming closer to understanding the laws of nature. Within the span of a couple of years, South Florida, where the University of Miami is located, has become the place for gathering of world class scientists including physicists, chemists, biologists, even economists, and political scientists who were

14

stimulated towards fundamental breakthroughs in their own fields. For example, in the CGC.III on Symmetry Principles at High Energy, there were some new comers who had already heard of the CGC.I and CGC.II. These included N. Cabibbo, Freeman Dyson, whom I referred to earlier, D. Frisch, T. Fulton, and Murray Gell-Man, who had not yet been awarded the Nobel prize, L. O' Raifeartaigh, M. Peshkin, J.C. Polkinghorne, N. Samios, W. Thirring, V. Weiskopf, and many others. Julian Schwinger and J. Robert Oppenheimer were among the participants. Sevda and I decided to organize a small dinner party, as in the previous years, in our home and invited Dr. and Mrs. Oppenheimer, Dr. and Mrs. Schwinger, Glenn Seaborg, Victor Weiskopf and Dr. and Mrs. Henry King Stanford. Right after dinner the rest of the conference participants arrived for cocktails. Including the local invitees we had to entertain nearly two hundred people. Luckily the backyard of the house could accommodate most of them. The guests began thinning out around midnight. The next morning session at 8:30 was fully attended.

Oppenheimer opened the conference and chaired the first session. Sessions II, III, and IV were chaired by Gregory Breit, Victor Weisskopf, and Walter Thirring, respectively. The first few Coral Gables Conferences were summarized perennially by Yuval Ne'eman of Tel Aviv University.

The next rencontre with Oppenheimer was on the occasion of the 1966 high energy physics conference at Berkeley, California. We discussed the organization of the 1967 conference. I also met Emilio Segre who had expressed a great interest in attending the CGC.IV in January 1967. While in Berkeley someone had shown me an Oppenheimer article which appeared in a book "Preludes in Theoretical Physics" edited by A. De-Shalit, H. Fesbhach, L. Von Hove, John Wiley and Sons, Inc. New York, 1966, where on page 75 he writes "More recently, attempts have been made to marry the exact symmetry of the poincaré group with broken internal symmetries. There are now adequate mathematical proofs that such structures are either trivial and that there is no marriage, or involve physical consequences wholly alien to our experience with the physics of particles; yet these efforts, largely by four brilliant Turks, may not, as a half millennium ago, in their turn lead to the discovery of America." Oppenheimer looked somewhat pale and fragile and had shown some concern regarding a possible absence from CGC.IV. His letter of January 12, 1967 to me only twelve days before the conference and four weeks before his death says it all--that Oppenheimer was terminally ill. He died on February 20, 1967 from throat cancer at a relatively young age of 62.

THE INSTITUTE FOR ADVANCED STUDY
PRINCETON, NEW JERSEY 08540

SCHOOL OF NATURAL SCIENCES

12 January 1967

Dear Professor Kursunoglu:

At this point, two things: Dr. Martin Halpern has been a member of the Institute this year after spending last year at Berkeley. He has found some quite interesting things on the interrelations of conservation laws, subtraction terms in dispersion relations and super-convergence conditions. He seems to be a very good physicist, whose interests have broadened and ripened well. He would like to come to the Coral Gables conference this year, and I recommend him to you for an invitation.

I have tried to keep an open mind about my own attendance, wanting very much, as you may guess, to come to the conference and spend some time at Coral Gables. I cannot now with any candor pretend that I can come, or even if I come, be of any use to anyone or any comfort. You should make your plans knowing that I cannot be there. You have my good wishes; and I should always be glad to talk with you of any problems that you would like to discuss.

With cordial greetings from both of us to both of you.

Robert Oppenheimer

Professor Behram Kursunoglu
Director
Center for Theoretical Studies
University of Miami
Coral Gables, Florida 33124

With the passing of J. Robert Oppenheimer, perhaps a phase of the nuclear era, the US leadership in guiding theoretical physics, a renaissance man, controversial affairs of national and international extent have come to an end. Not really! Over 25 years after his death, stories of Oppenheimer in the media, in books, in TV productions, and in Hollywood movies, are still continuing. His absence from the CGC.IV was a great disappointment to all of us. Julian and Clarice Schwinger had visited Oppenheimer at the Institute a few days before their arrival to Coral Gables. We were especially pleased when they told us that both Robert and Kitty Oppenheimer would have liked to see me and Sevda and had some kind words about us. Sevda and I had made plans to visit them after the conference. Unfortunately, it did not come to be, instead, I received a cable from Carl Kaysen, the new director of the Institute, inviting me to Oppenheimer's memorial ceremonies. Arnold Perlmutter and I went to Princeton on February 25, 1967 to attend the ceremonies. It was a very sad occasion. Marvin Goldberger, then chairman of the physics department of Princeton University, was acting as one of the ushers directing people to their seats in the auditorium. After the ceremonies we were invited to the new library of the Institute where Kitty Oppenheimer would receive guests. After attending the reception in the Library we returned the same day to Coral Gables.

February 20, 1967

TELEGRAM TO MRS. OPPENHEIMER:

SEVDA AND I WISH TO BE WITH YOU IN THIS MOMENT OF SADNESS FOR ALL WHO LOVE TRUTH AND WISDOM. ON BEHALF OF THE ENTIRE ACADEMIC COMMUNITY IN MIAMI I WISH TO CONVEY TO YOU OUR DEEPEST CONDOLENCES AND AFFECTION.

BEHRAM KURSUNOGLU

WIRE: Recvd. 12.55 p.m. Feb. 20

TO: DR. KURSUNOGLU

MEMORIAL SERVICE FOR ROBERT OPPENHEIMER SATURDAY, FEBRUARY 25, 3:00 P.M. ALEXANDER HALL, PRINCETON. PLEASE INFORM ASSOCIATES.

CARL KAYSEN
INSTITUTE ADVANCED STUDY

ROBERT OPPENHEIMER
1904 - 1967

Requiem Canticles 1965-1966   Igor Stravinsky
Recording conducted by Robert Craft

Hans Bethe
Henry DeWolf Smyth
George F. Kennan

Quartet Number 14 in C Sharp Minor, Op. 131
VI. Adagio; VII. Allegro   Ludwig van Beethoven
The Juilliard String Quartet

In the meantime, in view of Oppenheimer's eminent place in physics in American and in world society, I began thinking also of his role, by accepting membership in the CTS Scientific Council, in establishing CTS. We had many reasons to institute an annual memorial occasion. I thought of the possibility of an annual Oppenheimer memorial prize or award for achievements in theoretical physics, biology, and chemistry. I discussed the idea with some of the CGC participants. When I looked at the list of CGC.IV participants I could see quite a few potential candidates. The idea was approved by the Scientific Council of CTS and was released to the press.

"The Center for Theoretical Studies announces the establishment of a J. Robert Oppenheimer Memorial Prize for significant contributions to the fields of mathematics, theoretical physics, chemistry, biology, and to the philosophy of science. Nominations for this award should be addressed to the attention of the Scientific Council at the Center.

"The prize, which consists of $1,000, a Gold Medal, and a Citation, is to be a tribute to the memory of the late Professor Oppenheimer, as a former member of the Scientific Council, who was so instrumental in the founding of the Center. It is intended also to stimulate further work by the recipients, who are chosen for their promise for further achievements."

For several years the prize was sponsored by the late Alfred Sklar and his wife Olga Ferrer (M.D.). The Sklars have been most loyal supporters of CTS and have opened their lovely home to annual conference participants with lavish dinner receptions for over 200 guests at each time. Latin music, wine, and champagne were served with good food. Alfred Sklar was a

UNIVERSITY OF MIAMI
CORAL GABLES, FLORIDA 33124

December 11, 1967

Mrs. Robert Oppenheimer
Olden Farm
Princeton, New Jersey

Dear Kitty,

As a small tribute on our part to the great impact that the late Professor Robert Oppenheimer had on theoretical physics and on the growth of our Center, it has been proposed that we establish an annual prize in his name for scientists who have made outstanding contributions to the theoretical natural sciences. Dr. Alfred Sklar, a local resident and distinguished scientist in his own right, has agreed to provide annually a sum of $1,000 for the award.

This question will be brought to the agenda of the Scientific Council of the Center when it convenes in January if the proposal meets with your approval. It was this same Scientific Council that Dr. Oppenheimer served with his unmatchable brilliance for such an unhappily brief period.

I hope you are well and that you will forgive this intrusion. Best regards from Sevda and myself.

Sincerely yours,

Behram Kursunoglu

BK:cas

THE INSTITUTE FOR ADVANCED STUDY
PRINCETON, NEW JERSEY 08540

SCHOOL OF NATURAL SCIENCES

4 January 1968

Dear Behram:

I am most grateful that you propose to establish a prize at the Center in my husband's name for scientists who have made contributions to the theoretical natural sciences. I am especially pleased that it is to be an annual prize.

I hope that you and Sevda are well, and please thank Dr. Sklar for me for his generosity in providing the sums for the awards.

Best wishes for the New Year.

Sincerely,

Mrs. Robert Oppenheimer

Dr. Behram Kursunoglu
Center for Theoretical Studies
University of Miami
Coral Gables, Florida 33124

well-known theoretical chemist and had in 1938 authored a classic paper on Benzene Molecule with Edward Teller. Olga is a distinguished eye surgeon and is still taking care of her patients. Sklar had the habit of using the maximum power of his voice as an additional mode of persuasion during the presentation of his ideas as he stood by and wrote on the blackboard. On some occasions the high pitch of his voice was heard outside my office and the secretaries would run and open the door to see if all is well between me and Alfred!

Back to CGC.IV where the faces of the participants included J.D. Bjorken; N. Cabibbo; S. Fubini; F. Gürsey; E. Inönü; A. Martin; S. Okubo; F.M. Pipkin; B. Sakita; J.J. Sakurai; R.A. Schluter; S.S. Schweber; R.D. Tripp; O. Sinanoglu, who was at CTS on leave of absence from Yale University's Department of Chemistry; and Y. Yamaguchi. From the federal agencies the invitees included R.E. Denfeld, E. Weigold, and L.A. Wood from the US Air Force Office of Scientific Research and W. Gruner from the National Science Foundation.

With all these people attending, year after year, these "summit" meetings in high energy physics, it was time to carry this knowledge to a group of graduate students in a summer school or a Summer Institute. I had a share in the idea of initiating NATO-supported summer schools when I served as a member of the NATO science committee during 1957-58 representing the government of Turkey. I sent a proposal to NATO Science Committee, while it was located in the Paris headquarters of NATO, to organize a summer school in Istanbul, August 1-19, 1966. The lecturers, besides myself, were Sidney Colman (Harvard), Christain Fronsdal (UCLA), Asim O. Barut (Colorado), Yoichiro Nambu (Chicago), Leon Von Hove (CERN), Florian Scheck (Freiburg), Oktay Sinanoglu (Yale), Sergio Fubini (Torino), Sheldon L. Glashow (Harvard), Derek W. Robinson (CERN), David I. Olive (Cambridge). The 40 students were selected from various NATO member countries. The School was located at the beautiful Çinar Hotel by the Bosphorus strait, Istanbul. This seat of seven layers of civilizations over the past 5,000 years was conquered by the Ottoman Turks in 1453 thereby ending the Eastern Roman Empire, just about 40 years before the discovery of America which, in distant hindsight, was the dawn of the renaissance in Europe. The fall of Constantinople, so named as from A.D. 300-1930 may have had an impact, without Ottoman Turks direct participation, on the great revival of art, literature and learnings in Europe. In that era of transition from the medieval world to the modern, the Turks had seen their national security on the other sides of their frontiers, and under the banner of Islam in 1683 they reached the gates of Vienna. The City of Istanbul with its historic monuments, works of art, magnificent mosques with their architectural masterpieces displayed in their main domes, and gold topped minarets, water fountains, towers constitute a living history of the Ottoman Empire, one of the largest, most diverse in history. Therefore the NATO summer school in Istanbul was more than a series of lectures on high energy physics. It also provided exposure to the arts, literature, and architecture of a great civilization and allowed reflection on the history of mankind in a distant mirror. For example, the architect Sinan was not only one of the greatest in Ottoman history, but also one of the masters amongst the masters who left their marks in the great cities of Europe, and Asia.

SECTION II
NEUTRINO PHYSICS

NEUTRINO OSCILLATIONS AT ACCELERATORS

F. Vannucci

LPNHE, Univ. Paris 7
75252 Paris Cedex 05

INTRODUCTION

Hints of physics beyond the Standard Model come from solar[1] and atmospheric[2] neutrino experiments, where apparent deficits are interpreted as signs of oscillations, and also from the claim of the LSND experiment,[3] where a signal of $\bar{\nu}_e$ is observed in a beam which does not contain this flavour at production. Figure 1 shows the preferred regions corresponding to these three 'evidences'. They do not necessarily apply to the same physical oscillation channel.

What is then the place of short-baseline accelerator searches in the context of all these studies?

Accelerators offer clear advantages over natural sources of neutrinos. The beams are well-known in flux and composition. For example, high-energy beams produced from π and K decays are predominantly composed of ν_μ's with a small contamination of ν_e's. The flux is well controlled by detectors which measure accurately the accompanying muons in the shielding preceding the detectors.

Accelerator beams also offer advantages over lower energy beams. High enough energies allow the production of charge-current interactions of ν_μ's and even ν_τ's. Being above threshold permits appearance searches, namely the search for events due to neutrinos of a type absent in the original beam. The oscillation can, in principle, be proven on the basis of a unique event clearly identified. This gives a sensitivity much increased with respect to disappearance searches which are used in particular at reactors.

Thus, short-baseline experiments at accelerators remain the domain of high statistics searches, allowing to test small mixings, but they are limited in the parameter distance/energy which dictates the reach in the mass parameter δm^2, in practice not below about 10^{-1} eV^2.

High Energy Physics and Cosmology
Edited by B.N. Kursunoglu *et al.*, Plenum Press, New York, 1997

THE $\nu_\mu \, \nu_\tau$ CHANNEL

The present effort at accelerators focuses on the search for the oscillation $\nu_\mu \, \nu_\tau$ where short-baseline experiments are at an advantage. The interest of this channel comes from arguments of cosmology and astrophysics:

- the Big Bang model tells us that there are fossil neutrinos in the Universe at the level of $110/cm^3$ for each flavour. Neutrinos are about one billion times more abundant than nucleons, and a very small mass of each of them could dramatically change the gravitational equilibrium of the Universe.

- there is a missing mass, detected in particular at the level of dark halos in galaxies. This mass does not radiate but interacts only through gravitation. Present structure formation models favour mixed dark matter with a 20–30% hot component.[4] It is tempting to identify this component with the fossil neutrinos. This results in a mass relation:

$$\sum m_\nu = 10-20 \text{ eV} .$$

Assuming a hierarchy between the three neutrino masses, such a mass of about 10 eV would apply to the ν_τ. This is in agreement with the see-saw prescription which relates the ν_τ mass with a ν_μ mass of a few 10^{-3} eV suggested by the solar neutrino result. Is this a pure coincidence?

Such a high mass for the ν_τ gives a very comfortable oscillation length, commensurate with usual distances existing between production target and detectors at accelerators. The mixing to be expected is not predicted, but an improvement by one order of magnitude is a reason valid enough to repeat an experiment.

The challenge comes from the difficulty of detecting the τ^- lepton born in ν_τ interactions. It decays in about 1 mm at present energies. This problem is solved in e^+e^- experiments where the production vertex is well defined. In neutrino interactions the τ^- can arise anywhere inside a target which must have a large volume because of the size of the beam and the need of a mass high enough to generate events in sufficient quantity.

THE CERN PROGRAMME

Two experiments are currently taking data in the CERN high-energy neutrino beam, mainly composed of neutrinos of type ν_μ with average energies of 30 GeV. The experiments are searching for a signal of ν_τ in the mass region which is relevant for cosmology.

High spatial resolution in the whole volume of the target can be obtained with emulsions. This is the option chosen by the CHORUS Collaboration[5] which uses 800 kg of emulsions followed by a complete spectrometer/calorimeter essential to select a limited sample of candidates which then undergo the precise scanning.

About 500 000 events have been accumulated in 1994 and 1995, and part of the statistics has been scanned. Figure 2 shows one event with a short-lived particle reconstructed in the detector. A preliminary limit on the oscillation parameter has been presented recently[6]:

$$\sin^2 2\Theta \lesssim 8 \times 10^{-3} \text{ for masses such that } \delta m \gtrsim 30 \text{ eV}^2 .$$

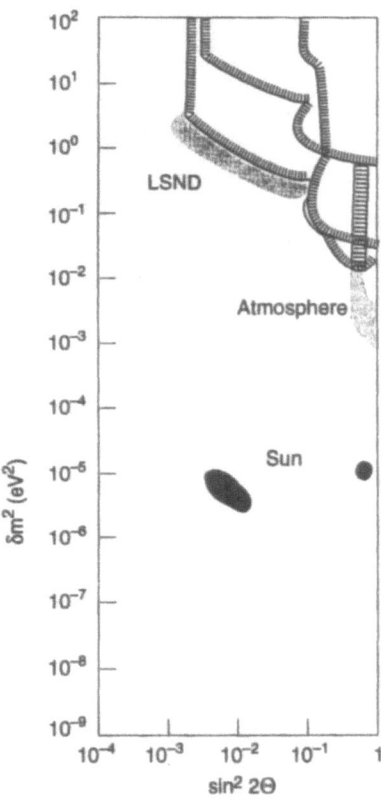

Figure 1. Currently favoured regions in oscillation searches for the three claims of the solar, atmospheric, and LSND solutions.

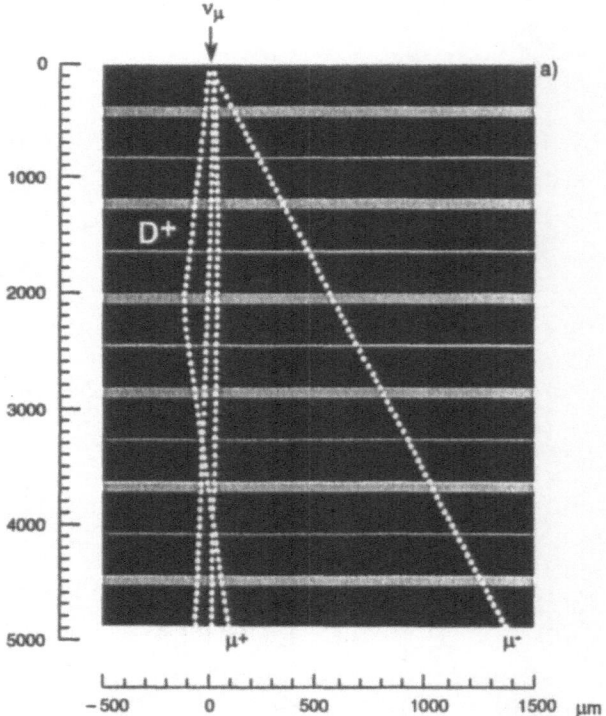

Figure 2. Event with a short-lived particle reconstructed in the CHORUS experiment.

NOMAD[7] has chosen a complementary approach. The detector resembles an 'electronic bubble chamber' with a target made of low-Z material to allow a very good precision of reconstruction and to limit hadron reinteractions and photon conversions. Moreover specialised detectors give an accurate identification of e, γ and μ. The ν_τ signal is extracted by kinematical criteria.

The power of the reconstruction is demonstated in Figure 3 where the K^0 invariant mass is shown. The preliminary result is slightly better than the published one[8]:

$$\sin^2 2\Theta \lesssim 4 \times 10^{-3}.$$

Both experiments will take more data this year, and they hope to reach the level of a few 10^{-4} when the analyses are completed.

PERSPECTIVES

At this point, it is useful to assemble all the information obtained so far in oscillation searches together with some prejudices that one can have on massive neutrinos:

a) solar neutrinos indicate the disappearance of ν_e with $\delta m^2 \simeq 10^{-5}$ eV^2;

b) atmospheric neutrinos point to the disappearance of ν_μ with $\delta m^2 \simeq 10^{-2}$ eV^2;

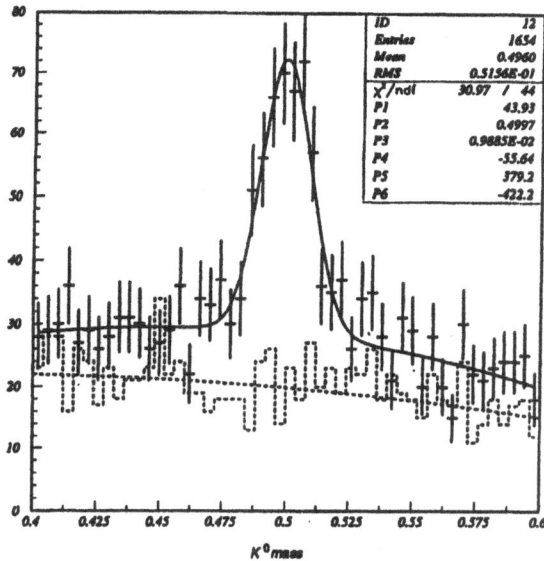

Figure 3. Invariant mass showing a signal of K^0 in the NOMAD detector.

c) LSND claims evidence for the oscillation $\bar{\nu}_\mu \bar{\nu}_e$, equivalent to $\nu_\mu \nu_e$ insofar as CP is conserved with $\delta m^2 \simeq 1$ or 2 eV^2;

d) to play a role in the missing mass of the Universe, the neutrino masses must fulfil the condition $\sum m_\nu \simeq 10$ eV;

e) the see-saw recipe suggests a strong hierarchy between masses $m(\nu_e) \ll m(\nu_\mu) \ll m(\nu_\tau)$.

Clearly these five propositions are incompatible if we limit ourselves to the three known neutrinos having only electroweak interactions.

Let us first consider propositions a) and c) which both test the $\nu_e \nu_\mu$ channel. They suggest very different δm^2. Nevertheless they are compatible if one considers the phenomenology of oscillations among three neutrinos.[9] The LSND result could signal the 'indirect' oscillation between ν_e and ν_μ governed by the ν_τ mass and corresponding to a small mixing. On the other hand, solar neutrinos which correspond to a large distance/energy parameter are sensitive to the term governed by a small mass term and a relatively large mixing.

Having demonstrated the compatibility of a) and c), it is simple to see that four out of the five propositions are consistent, the only one to be dismissed being b).

In this scheme the neutrino masses are given by:

$$m(\nu_\tau) \simeq 1 \text{ eV}, \; m(\nu_\mu) \simeq 10^{-3} \text{ eV}, \; m(\nu_e) \simeq 10^{-6} \text{ eV} .$$

There is an alternative to this solution which is obtained if one wants to retain both solar and atmospheric hints. The LSND result has to be discarded. We are then left with three neutrino masses and two different δm^2. In order to fulfil the cosmological requirement, one has to relinquish the mass hierarchy and the obvious solution gives three mass-degenerate neutrinos of a few eV each. This is not the preferred solution for accelerator searches since it leaves the short-baseline experiments with no hope.

Several other variants exist which lead to different prescription! After thirty years of study, neutrinos remain the most unknown of the known particles. Many questions are left without clear answers concerning their masses, mixings, particle/antiparticle properties, magnetic moment, etc. Furthermore, the experimental front is always alive with new puzzles, the last one being the LSND claim. Let us hope that oscillations will be firmly demonstrated soon, and that the long history of neutrino masses will find a definitive conclusion.

REFERENCES

1. GALLEX Collab., *Phys. Lett.* B327: 377 (1994); SAGE Collab., *Phys. Lett.* B328: 234 (1994); Homestake Collab., *Nucl. Phys.* B38: 47 (1995); Kamiokande Collab., ibid, 55.
2. Kamiokande Collab., *Phys. Lett.* B335: 237 (1994); IMB Collab. *Phys. Rev.* D46: 3720 (1992).
3. A. Athanassopoulos et al., *Phys. Rev.* C54: 2685 (1996).
4. E.L. Wright et al., *Astrophys. J.* 396: L13 (1992); J.A. Holtzman and J.R. Primack, *Astrophys. J.* 405: 428 (1993).
5. CHORUS Collaboration, N. Armenise et al., CERN-SPSLC/90–42.
6. Status reports, Meeting of the CERN/SPSLC, January 21, 1997.
7. NOMAD Collaboration, P. Astier et al., CERN-SPSLC/91–48.
8. N. Ushida et al., *Phys. Rev. Lett.* 57: 2897 (1986).
9. C.W. Kim and H. Nishiura, Preprint JHU-HET 8506 (1985); S.M. Bilenky et al., Preprint DFTT 25/95.

KARMEN: PRESENT NEUTRINO-OSCILLATION LIMITS AND PERSPECTIVES AFTER THE UPGRADE

Guido Drexlin[a]

Porschungszentrum Karlsruhe
Institut für Kernphysik I
D-76021 Karlsruhe, Postfach 3640, Germany
E-Mail: guido@ik1.fzk.de

ABSTRACT

The neutrino experiment KARMEN is situated at the beam stop neutrino source ISIS. It provides ν_μ's, ν_e's and $\bar{\nu}_\mu$'s in equal intensities from the $\pi^+-\mu^+$-decay at rest. The oscillation channels $\nu_\mu \to \nu_e$ and $\bar{\nu}_\mu \to \bar{\nu}_e$ are investigated with a 56 t liquid scintillation calorimeter at a mean distance of 17.6 m from the ν-source. No evidence for oscillations could be found with KARMEN, resulting in 90% CL exclusion limits of $sin^2(2\Theta) < 8.5 \cdot 10^{-3}$ ($\bar{\nu}_\mu \to \bar{\nu}_e$) and $sin^2(2\Theta) < 4.0 \cdot 10^{-2}$ ($\nu_\mu \to \nu_e$) for $\Delta m^2 > 100$ eV^2.

In 1996, the KARMEN neutrino experiment has been upgraded by an additional veto system. Vetoing of cosmic muons passing the 7000 t massive iron shielding of the detector suppresses energetic neutrons from deep inelastic scattering of muons as well as from μ-capture in iron. Up to 1996, these neutrons penetrating into the detector represented the main background for the $\bar{\nu}_\mu \to \bar{\nu}_e$ oscillation search. With an expected reduction of the background rate by a factor of 40 the experimental sensitivity for $\bar{\nu}_\mu \to \bar{\nu}_e$ will be significantly enhanced towards $sin^2(2\Theta) \approx 1 \cdot 10^{-3}$ for large Δm^2.

1. Neutrino Production at ISIS

The Karlsruhe Rutherford Medium Energy Neutrino experiment KARMEN is being performed at the neutron spallation facility ISIS of the Rutherford Appleton Laboratory. The neutrinos are produced by stopping 800 MeV protons in a beam dump Ta-D_2O-target. Neutrinos emerge from the consecutive decay sequence $\pi^+ \to \mu^+ + \nu_\mu$ and $\mu^+ \to e^+ + \nu_e + \bar{\nu}_\mu$.

---

[a]for the KARMEN Collaboration

Thus, ISIS represents a ν-source with identical intensities for ν_μ, ν_e and ν_μ emitted isotropically ($\Phi_\nu = 6.37 \cdot 10^{13}\ \nu/s$ per flavor for p-beam current $I_p = 200\ \mu A$). The energy spectra of the ν's are well defined due to the decay at rest of both the π^+ and μ^+ (fig. 1a). The ν_μ's from

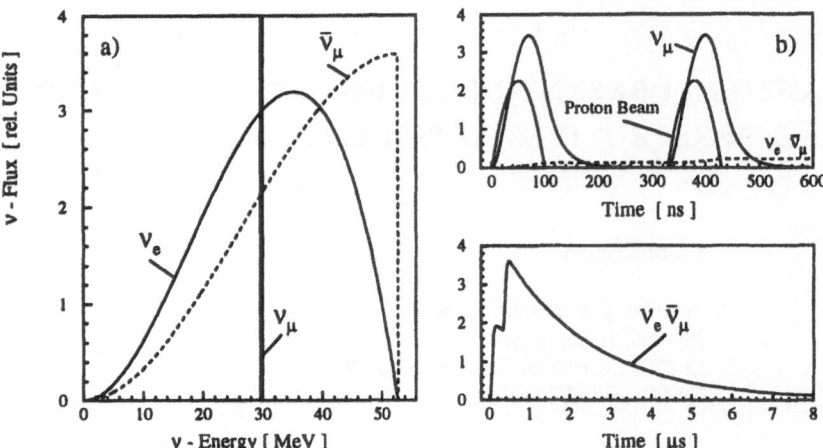

Fig. 1. Neutrino energy spectra (a) and production times (b) at ISIS.

π^+–decay are monoenergetic ($E_\nu = 30$ MeV), the continous energy distributions of ν_e and $\bar{\nu}_\mu$ up to 52.8 MeV can be calculated using the V–A theory. Since π^+ and μ^+ are stopped in the small beam dump target, the ν production region is essentially limited to a volume of ± 5 cm radial to the proton beam and ± 10 cm along the beam axis. With a mean distance source–detector of $L = 17.6$ m and including the spatial resolution of the detector, the uncertainty $\Delta L/L$ for the ν flight path is less than 1%. ISIS therefore ensures that the important experimental parameters L and E_ν for ν–oscillations are determined with high precision.

Two parabolic proton pulses of 100 ns basis width and a gap of 225 ns are produced with a repetition frequency of 50 Hz (fig. 1b). The different lifetimes of pions ($\tau = 26$ ns) and muons ($\tau = 2.2\ \mu s$) allow a clear separation in time of the ν_μ-burst from the following ν_e's and $\bar{\nu}_\mu$'s. Furthermore the accelerator duty cycle allows effective suppression of any beam-uncorrelated background by four to five orders of magnitude.

2. The KARMEN Detector

The neutrinos are detected in a 56 t liquid scintillation calorimeter [1]. A massive blockhouse of 7000 t of steel in combination with a system of two layers of active veto counters provides shielding against beam correlated spallation neutron background, suppression of the hadronic component of cosmic radiation as well as reduction of the flux of cosmic muons. The central scintillation calorimeter and the inner veto counters are segmented by double acrylic walls with an air gap allowing efficient light transport via total internal reflection of the scintillation light at the module walls. The event position is determined by the individual module and the time difference of the PM signals at each end of this module. Due to the optimized optical properties of the organic liquid scintillator and an active volume of 96% for the calorimeter, an excellent energy resolution of $\sigma_E = \frac{11.5\%}{\sqrt{E[MeV]}}$ is achieved. In addition, Gd_2O_3 coated paper within the module walls provides efficient detection of thermal neutrons due to the very high capture

limit of the oscillation probability P of

$$P_{\nu_\mu \to \nu_e} \; < 3.8/187.3 = 2.0 \cdot 10^{-\cdot} \qquad (90\% \, CL\,)$$

can be extracted. Due to the normalization of the full oscillation expectation this result is very reliable. The oscillation search is nearly background free so that the sensitivity for this oscillation channel is essentially limited by the relatively small expectation value for full oscillation, i.e. statistics.

4. Oscillation limits on $\bar{\nu}_\mu \to \bar{\nu}_e$

The most sensitive mode of the KARMEN experiment for the search of ν-oscillations is the $\bar{\nu}_\mu \to \bar{\nu}_e$ channel. First, $\bar{\nu}_e$'s are not produced within the ISIS target apart from a very small contamination of $\bar{\nu}_e/\nu_e < 6 \cdot 10^{-4}$. The detection of $\bar{\nu}_e$'s via $p\,(\,\bar{\nu}_e\,,e^+\,)\,n$ would therefore indicate oscillations $\bar{\nu}_\mu \to \bar{\nu}_e$ in the appearance channel. Secondly, the cross section for $p\,(\,\bar{\nu}_e\,,e^+\,)\,n$ with $\bar{\nu}_e$'s from oscillations ($\sigma = 93.6 \cdot 10^{-42}\mathrm{cm}^2$ for $\Delta m^2 = 100\,\mathrm{eV}^2$) is about 20 times larger than for $^{12}\mathrm{C}\,(\,\nu_e\,,e^-\,)\,^{12}\mathrm{N_{g.s.}}$ with ν_e's from $\nu_\mu \to \nu_e$ ($\sigma = 4.95 \cdot 10^{-42}\mathrm{cm}^2$), and the ratio of target nuclei of the scintillator is $H/C = 1.767$.

The signature for the detection of $\bar{\nu}_e$'s is a spatially correlated delayed coincidence of positrons from $p\,(\,\bar{\nu}_e\,,e^+\,)\,n$ with energies up to $E_{e^+} = E_{\bar{\nu}_e} - Q = 52.8 - 1.8 = 51\,\mathrm{MeV}$ (fig. 3a) and γ emission of either of the two neutron capture processes $p\,(\mathrm{n},\gamma)\,\mathrm{d}$ or $\mathrm{Gd}\,(\mathrm{n},\gamma)\,\mathrm{Gd}$ with γ energies of 2.2 MeV or up to 8 MeV, respectively (fig. 3b). The positrons are expected in a time

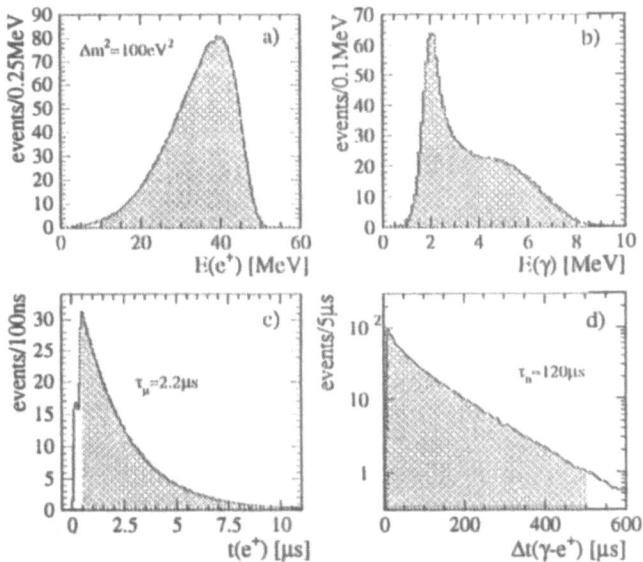

Fig. 3. Expected signature for $\bar{\nu}_\mu \to \bar{\nu}_e$ full oscillation:
a) MC energy of prompt positron for $\Delta m^2 = 100\,\mathrm{eV}^2$; b) energy of sequential γ's; c) time of prompt event relative to ISIS beam-on-target; d) time difference between prompt e^+ and sequential γ's; shaded areas are accepted by evaluation cuts.

window of 0.5 to 10.5 μs after beam-on-target (fig. 3c). The neutrons from $p\,(\,\bar{\nu}_e\,,e^+\,)\,n$ are thermalized and captured typically with $\tau = 120\,\mu$s (fig. 3d). The neutron detection efficiency for the analyzed data is 28.2%. The data set remaining after applying all cuts in energy, time and spatial correlation is shown in fig. 4. A prebeam analysis of cosmic ray induced sequences

cross section of the Gd(n,γ) reaction. The KARMEN electronics is synchronized to the ISIS proton pulses to an accuracy of better than $\pm 2\,\text{ns}$, so that the time structure of the neutrinos, especially of ν_μ's, can be exploited in full detail. In these proceedings we only present results of the oscillation search and its results obtained in the first data taking period from 1990–1995, the investigation of ν–nucleus interactions on ^{12}C is described elsewhere[2] in detail.

3. Oscillation limits on $\nu_\mu \to \nu_e$

In the event of $\nu_\mu \to \nu_e$ oscillations, monoenergetic 30 MeV ν_e's would arise in the ν_μ time window after beam-on-target. The ν_e detection reaction is $^{12}\text{C}(\nu_e, e^-)\,^{12}\text{N}_{\text{g.s.}}$ followed by the β–decay $^{12}\text{N}_{\text{g.s.}} \to\,^{12}\text{C} + e^+ + \nu_e$. One would therefore expect electrons with a peaked energy spectrum ($E_{e^-} = E_\nu - Q = 29.8 - 17.3 = 12.5\,\text{MeV}$, see fig. 2a) within the two ν_μ time pulses (fig. 2b). This characteristic prompt energy signal is followed by the energy of the sequential spatially correlated e^+ (fig. 2c) which follows the e^- with the typical ^{12}N decay time (fig. 2d). In

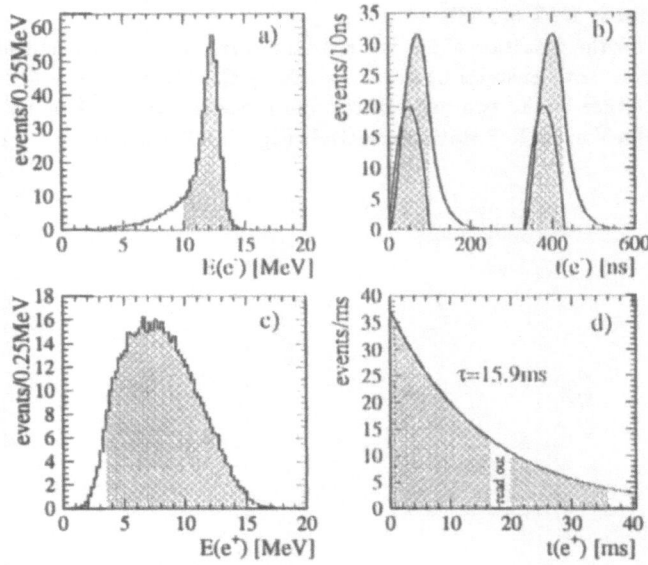

Fig. 2. Expected signature for $\nu_\mu \to \nu_e$ full oscillation:
a) simulated MC energy of prompt event; b) proton pulses and time of prompt event relative to ISIS beam-on-target; c) MC energy of sequential event; d) time difference between prompt and sequential event; shaded areas show the allowed regions of evaluation cuts.

the later ν_μ time window one can measure the number of $^{12}\text{C}(\nu_e, e^-)\,^{12}\text{N}_{\text{g.s.}}$ reactions induced by ν_e's from μ^+-decay in the ISIS target, which can be used to calculate the expectation of $\nu_\mu \to \nu_e$ induced CC reactions for $P_{\nu_\mu \to \nu_e} = 100\%$. The different detection efficiencies and the energy dependence of the cross section have to be taken into account to extract the ν-flux and cross section independent expectation of 187.8 oscillation signatures. Applying all cuts (e.g. $10 \le E_{pr} \le 14\,\text{MeV}$; $0 \le t_{pr} \le 100\,\text{ns}$ or $325 \le t_{pr} \le 425\,\text{ns}$) only 2 sequences remain within the data taken between July 1992 and December 1995. 0.50 ± 0.20 cosmic induced events contribute to the background which is dominated by the small contribution of ν_e's from μ^+–decay within the two 100 ns long ν_μ-time intervals after beam-on-target (1.76 ± 0.2). With a total background of $N_{bg} = 2.26 \pm 0.3$ events, there is no hint for $\nu_\mu \to \nu_e$ oscillations and an upper

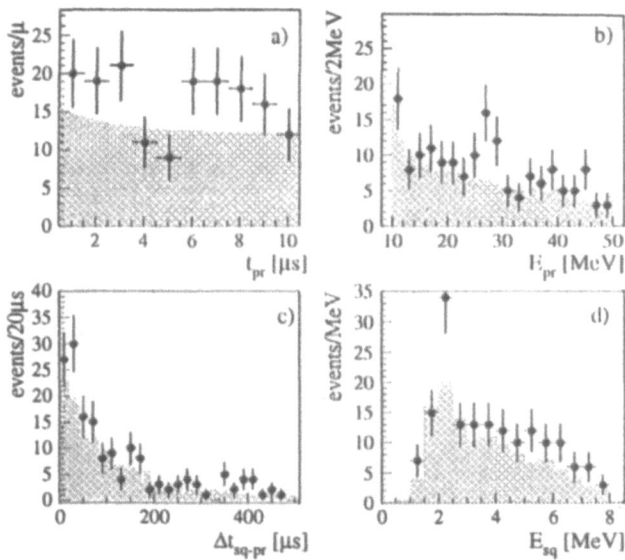

Fig. 4. Time (a,c) and energy (b,d) distribution of reduced sequences; shaded lines and histograms represent the pre-beam background (12.2 events per μs) plus ν_e-induced CC events and $\bar{\nu}_e$-contamination.

results in an accumulated background level of 12.2 ± 0.2 events per μs in the prompt $10\,\mu s$–window (see fig. 4a). The actual rate is $16.4\pm1.3/\mu s$ which corresponds to a beam excess of $2.4\,\sigma$ compared with the prebeam level including ν_e-induced CC (9 events) and $\bar{\nu}_e$-contamination (1.7 events). Although the secondary part of the sequences shows the typical signature of thermal neutron capture, the prompt time and energy distribution does not follow the expectation from $\bar{\nu}_\mu \rightarrow \bar{\nu}_e$ oscillation with $\Delta m^2 = 100\,\mathrm{eV}^2$.

To extract a possible small contribution of $\bar{\nu}_\mu \rightarrow \bar{\nu}_e$, the data set is scanned with a two-dimensional maximum likelihood analysis on time and energy distribution of the positrons requiring a $2.2\,\mu s$ exponential time constant for the e^+ and a time independent cosmic induced background. The measurement of the e^+ energy with spectroscopic quality is highly sensitive to changes in the energy spectrum due to the dependence of the oscillation probability on the mass term Δm^2 (see fig. 5a). The energy distributions of the positrons used in the likelihood analysis therefore have been tested with spectra for Δm^2 in the range from 0.01 to $100\,\mathrm{eV}^2$.

For most of the investigated parameter range of Δm^2 the likelihood analysis results in best fit values compatible with a zero signal within a 1σ error band (see fig. 5b). Only for a parameter region at $\Delta m^2 = 6.2\,\mathrm{eV}^2$ there is a positive signal 2.3σ above zero which is not considered as statistically significant. In addition, this Δm^2-value corresponds to the first theoretical oscillation minimum in the detector with the lowest possible mean energy of the positrons (see also fig. 5a) and represents therefore an extremum in the likelihood analysis which should be interpreted with special precaution.

On this basis of no evidence for oscillations, $90\%\,CL$ upper limits for oscillation events (fig. 5b) as well as for the oscillation parameters Δm^2 and $sin^2(2\Theta)$ are deduced. Our result can be compared with an expected signal of 6 up to 76 events for $\Delta m^2 = 3.9\,\mathrm{eV}^2$ based on a recently published oscillation probability of $P_{\bar{\nu}_\mu \rightarrow \bar{\nu}_e} = 0.0031$ by LSND [3]. For large Δm^2 we expect at KARMEN 1898 detected oscillation events for full oscillation which results in an upper limit of the mixing angle

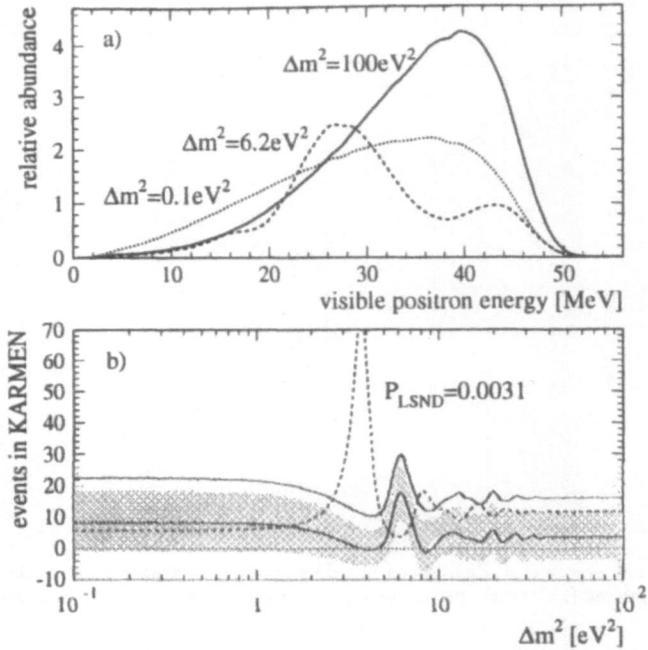

Fig. 5. a) Examples of expected e^+–spectra (visible energy including detector response) for different oscillation parameters Δm^2; b) likelihood-fit results depending on Δm^2; the shaded area represents the 1σ–error band around the best fit values for $\Delta m^2 = 0.01\ldots 100\,\mathrm{eV}^2$; the solid line above the shaded band shows the $90\%\,CL$ upper limit from KARMEN, the broken line the expected event numbers in KARMEN based on the LSND oscillation evidence [3].

$$sin^2(2\Theta) < 16.3/1898 = 8.5 \cdot 10^{-3} \quad \text{for large } \Delta m^2 \qquad (90\%\,CL\ ).$$

Fig. 6 shows the KARMEN exclusion curves in the parameter space of Δm^2 and $sin^2(2\Theta)$ in a two neutrino flavor oscillation calculation for the appearance channels $\nu_\mu \to \nu_e$ and $\bar\nu_\mu \to \bar\nu_e$ in comparison with other results of ν–oscillation searches at accelerators [4] and reactors [5]. As the sensitivity for $\bar\nu_\mu \to \bar\nu_e$ of the KARMEN experiment is comparable to that of LSND (both experiments expect about 2000 oscillation events for $\Delta m^2 > 100\,\mathrm{eV}^2$ and $sin^2(2\Theta) = 1$ on their data sample until 1995), the KARMEN $90\%\,CL$ exclusion curve cannot exclude the entire parameter space favoured by the positive result of LSND.

5. The KARMEN Upgrade

For at least the next 3 years, no other oscillation experiment will cover the whole parameter region favored by LSND. Only the running KARMEN experiment, with an improved sensitivity, will be able to crosscheck the evidence postulated by LSND. We therefore investigated different scenarios for increasing the $\bar\nu_\mu \to \bar\nu_e$ sensitivity. Whereas the sensitivity for $\nu_\mu \to \nu_e$ oscillations is essentially limited by statistics, the KARMEN sensitivity in the $\bar\nu_\mu \to \bar\nu_e$ channel can only be substantially increased by the reduction of the small but dominant cosmogenic background (see fig. 4). This background is induced by cosmic muons stopping or undergoing deep inelastic scattering in the iron shielding which surrounds the KARMEN detector and veto system. Energetic neutrons emitted in these processes can penetrate deep into the detector without

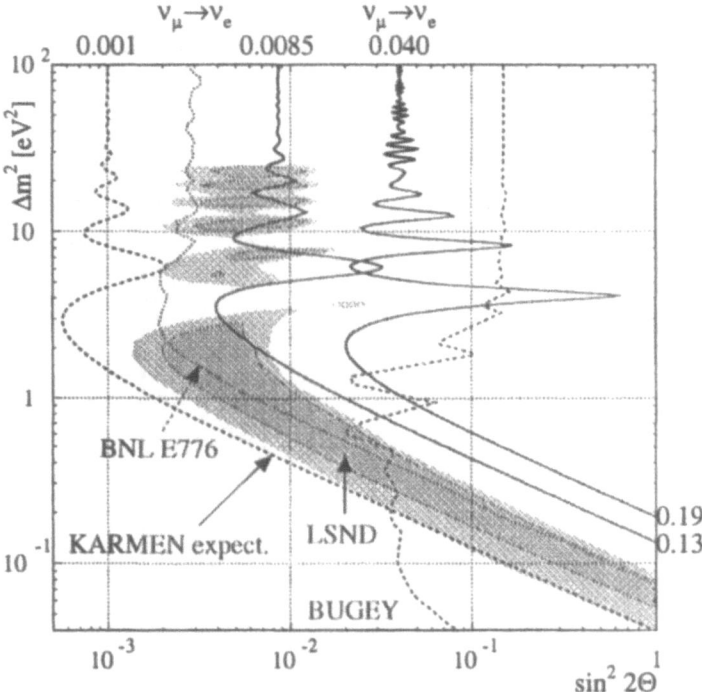

Fig. 6. 90% CL exclusion curves and limits for $\Delta m^2 = 100\,\mathrm{eV}^2$, $sin^2(2\Theta) = 1$ from KARMEN for $\nu_\mu \to \nu_e$ and $\bar{\nu}_\mu \to \bar{\nu}_e$ as well as the expected sensitivity for $\bar{\nu}_\mu \to \bar{\nu}_e$ after the upgrade; oscillation limits from BNL E776 [4] and Bugey [5]; LSND evidence is shown as shaded areas (90% CL and 99% CL areas respectively).

triggering the veto system, thus producing an event sequence of prompt recoil protons followed by the capture of the then thermalized neutrons.

To tag the original muons in the vicinity of the detector, a further active veto layer within the blockhouse has been built in 1996, which consists of 136 plastic scintillator bars (BICRON BC 412) of lengths up to 4 m, 65 cm width and 5 cm thickness, with a total surface of 300 m² covering all sides of the detector (fig. 7). There is at least 1 m between the new counter and the existing shield so that energetic neutrons produced by cosmic muons outside the new veto system have to travel a path of more than 4 attenuation lengths in iron ($\Lambda = 21\,\mathrm{cm}$). This reduces cosmogenic neutrons to a negligible fraction of less than 1.5% of the original flux. The new veto system is designed to reduce cosmogenic sequential background by a factor of at least 40. This reduction factor is based on detailed background measurements and extensive GEANT MC simulations of cosmic muons. First preliminary evaluations of cosmic background in a prebeam window indicate the expected reduction factor in the energy region of interest when the information of veto hits is included in the analysis.

After two years of measuring time with the new detector configuration, the KARMEN sensitivity for $\bar{\nu}_\mu \to \bar{\nu}_e$ is expected to exclude the whole parameter region of evidence suggested by LSND if no oscillation signal will be found (fig. 6). In that case, mixing angles with $sin^2(2\Theta) > 1 \cdot 10^{-3}$ will be excluded for large Δm^2. The veto upgrade will also increase the signal to background ratio of single prong ν-induced events on ^{12}C and therefore improve the

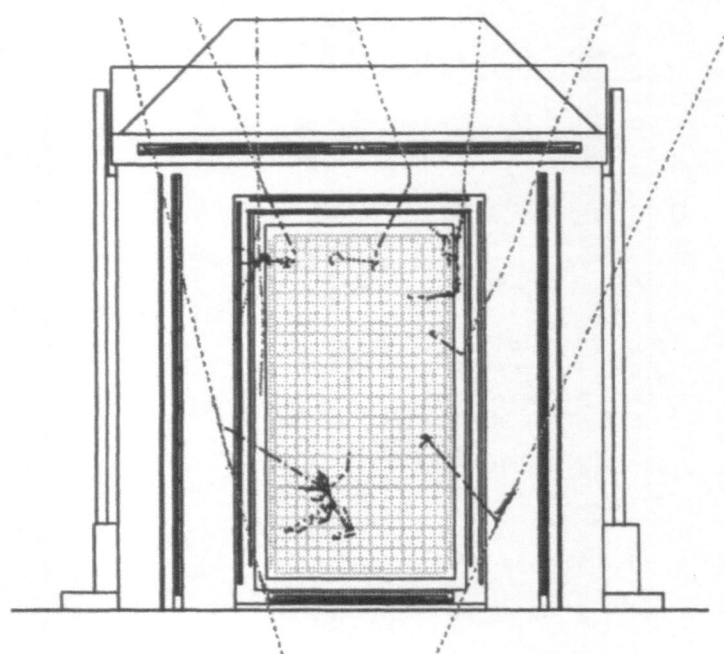

Fig. 7. Cross section of the KARMEN central detector with surrounding shield counters and massive iron blockhouse, now including the additional veto system. Cosmic muons passing or stopping in the iron which produce energetic neutrons can now be tagged by the new veto scintillators.

investigation of the published anomaly in this time distribution [6].

6. Acknowledgements

We acknowledge the financial support of the German Bundesministerium für Bildung, Wissenschaft, Forschung und Technologie.

7. References

1) G. Drexlin et al., *Nucl. Instr. and Meth.* **A289** (1990) 490.
2) B. A. Bodmann et al., *Phys. Lett.* **B339** (1994) 215.
   B. Zeitnitz et al.,*Prog. Part. Nucl. Phys.* **32** (1994) 351.
3) C. Athanassopoulos et al., *Phys. Rev. C* **54** (1996) 2685.
   C. Athanassopoulos et al., *Phys. Rev. Letters* **77** (1996) 3082.
4) L. Borodovsky et al., *Phys. Rev. Letters* **68** (1992) 274.
5) B. Achkar et al., *Nucl. Phys.* **B434** (1995) 503.
6) B. Armbruster et al., *Phys. Lett.* **B348** (1995) 19.

NEUTRINO AND ANTI-NEUTRINO REACTIONS
IN NUCLEI

S.L. Mintz, G.M. Gerstner, and M.A. Barnett

Physics Department
Florida International University
Miami, Florida 33199

M. Pourkaviani

MP Consulting Associates
714 Fox Valley Dr.
Longwood, Florida 32701

INTRODUCTION

The last year or two in neutrino-nuclei interactions has been spent in expectant and hopeful waiting rather than in striking new results. Perhaps the most interesting results have been from LAMPF for the inclusive muon neutrino reaction,$\nu_\mu + {}^{12}C \longrightarrow X + \mu^-$, and for the corresponding exclusive reaction $\nu_\mu + {}^{12}C \longrightarrow \mu^- + {}^{12}N_{(g.s.)}$. Here the relatively low cross section for the inclusive reaction has persisted and their measurement of the exclusive cross section is in good agreement with theory thus putting to rest some questions which have been raised. We shall discuss this in more detail later in this paper. The KARMEN collaboration continues to obtain data and their results for both inclusive and exclusive electron neutrino reactions on ${}^{12}C$ are in reasonable agreement with theory. New data from kamiokande II is beginning to appear but there is no definitive result yet. Finally the Sudbury Neutrino Observatory (SNO) is in its final phase of construction. We will have occasion to mention some of these experiments later.

In this paper we shall, in addition to work on ${}^{12}C$, also consider the possibility of observing both neutrino and anti-neutrino reactions in the ${}^3H$-${}^3He$ system. This system is interesting for a number of reasons. First transitions in this system are mirror transitions so that transition rates are relatively large. Secondly the percentage of neutrons in ${}^3H$ and protons in ${}^3He$ are larger than in any other nuclei which would be suitable targets for neutrino and anti-neutrino reactions respectively. This also tends to raise the transition rate. In addition the relatively large size of the axial current form factor F_A contributes to an enhanced transition matrix element. As we shall see, certain interference terms in the transition matrix element squared for these reactions are large. This also contributes to the interest in these processes.

In subsequent sections of this paper we shall discuss the present experimental situation with particular emphasis on the case of neutrino reactions in ^{12}C. We shall then focus on the neutrino and anti-neutrino reactions in ^{3}H and ^{3}He respectively before summing up the current situation.

PRESENT EXPERIMENTAL RESULTS

The KARMEN and LAMPF groups have been taking data on various neutrino processes for many years now and these efforts are well familiar to everyone active in this field. The longest term results have been on the ^{12}C nucleus. The first group to perform electron neutrino measurements on this system was the LAMPF group[1]. They were able to perform experiments for the exclusive charged current reaction, $\nu_e + {}^{12}C \longrightarrow e^- + {}^{12}N_{(g.s.)}$, and for the inclusive charged current reaction $\nu_e + {}^{12}C \longrightarrow e^- + X$. For the exclusive case they obtained a cross section:

$$< \sigma >= 1.04 \pm 0.10 \; (stat) \pm 0.10 \; (sys) \times 10^{-41} \; cm^2 \tag{1}$$

where the above average is over the Michel spectrum. The KARMEN[2] group has also measured this cross section and obtains:

$$< \sigma >= .93 \pm .08 \; (stat) \pm .075 \; (sys) \times 10^{-41} \; cm^2. \tag{2}$$

These results are in reasonable agreement with theoretical calculations[3,4,5] which may be written as:

$$< \sigma >\simeq .8 \sim .95 \times 10^{-41} \; cm^2. \tag{3}$$

There was no agreement between the two groups concerning the inclusive electron neutrino reaction. This is usually given as the cross section for the process, $\nu_e + {}^{12}C \longrightarrow e^- + {}^{12}N^*$, and the LAMPF results for this process were:

$$< \sigma >= 3.6 \pm 20\% \times 10^{-42} \; cm^2 \tag{4}$$

where the results are again averaged over the Michel spectrum. At the time of this measurement[1] there was only one calculation[3] which yielded:

$$< \sigma >= 3.7 \times 10^{-42} \; cm^2 \tag{5}$$

so that theory and experiment appeared to be in agreement. This same reaction was measured by the KARMEN and at the time they obtained a result greater than that of LAMPF by almost a factor of four. KARMEN's present value is:

$$< \sigma >= 6.4 \pm 1.3 \; (stat) \pm 1.2 \; (sys) \times 10^{-42} \; cm^2 \tag{5}$$

which may be compared to recent theoretical[6,7] calculations of:

$$< \sigma >\simeq 6.3 \times 10^{-42} \; cm^2 \tag{6a}$$

and

$$< \sigma > \simeq 7.75 \times 10^{-42} \ cm^2. \tag{6b}$$

The two theoretical claculations were by entirely different methods and considering the errors involved it may be said that there is some agreement between theory and experiment in the range of the Michel spectrum, although the LAMPF result is not fully understood.

It should be remarked that the KARMEN collaboration is also able to obtain a cross section for the neutral current process, $\nu_e + {}^{12}C \longrightarrow \nu_e + {}^{12}C^*$. Their result is presently:

$$< \sigma > \simeq 10.5 \pm 2.3 \ (stat) \pm 1.3 \ (sys) \times 10^{-42} \ cm^2 \tag{7}$$

which is in good agreement with theoretical predictions[5,8,9]:

$$< \sigma > \simeq 9.9 \pm 1.0 \times 10^{-42} \ cm^2. \tag{8}$$

This again gives us confidence in our understanding of exclusive neutrino reactions in ${}^{12}C$.

More recently the LAMPF group has obtained results for the inclusive process, $\nu_\mu + {}^{12}C \longrightarrow X + \mu^-$. Here the neutrinos are produced by muons decaying in flight and the spectrum is a higher energy one than that for the Michel spectrum. This spectrum has a broad peak from about 40 to 80 MeV and falls off to zero at around 280 MeV. Their results are in the range of:

$$< \sigma > \simeq 12.0 \pm 0.7 \ (stat) \pm 1.6 \ (sys) \times 10^{-40} \ cm^2. \tag{8}$$

These results were lower than was expected by two theoretical calculations. One making use of a Fermi gas model[10] yielded:

$$< \sigma > \simeq 24 \times 10^{-40} \ cm^2 \tag{9a}$$

and the other based on a random phase approximation[6] yielded:

$$< \sigma > \simeq 20 \times 10^{-40} \ cm^2. \tag{9b}$$

However their results were in accord with two other models by Mintz and Pourkaviani[11] and Kim and Mintz[12] respectively, the former yielding:

$$< \sigma > \simeq 13.1 \times 10^{-40} \ cm^2 \tag{10a}$$

and the latter yielding:

$$< \sigma > \simeq 13.3 \times 10^{-40} \ cm^2. \tag{10b}$$

These results have all been discussed in the literature and we shall not treat them further here.

However, at the time the LAMPF results were announced it was suggested that the neutrino spectrum might not be well known or that the experiment might suffer from other unspecified problems. It was therefore suggested that LAMPF might measure the exclusive cross section for the reaction, $\nu_\mu + {}^{12}C \longrightarrow \mu^- + {}^{12}N_{(g.s.)}$. They have done so and it is this process which we now discuss.

THE EXCLUSIVE MUON NEUTRINO REACTION IN ^{12}C

Although as we have just noted, the inclusive muon neutrino reaction is not an easy one to calculate, the same cannot be said for the exclusive muon neutrino reaction in ^{12}C, $\nu_\mu + {}^{12}C \longrightarrow \mu^- + {}^{12}N_{(g.s.)}$. A closely related reaction, $\nu_e + {}^{12}C \longrightarrow e^- + {}^{12}N_{(g.s.)}$, has been studied both experimentally[1,2], and theoretically[3,4,5] with reasonable agreement both among the experiments and among the theoretical calculations. We therefore present here a calculation of the cross section for the reaction, $\nu_\mu + {}^{12}C \longrightarrow \mu^- + {}^{12}N_{(g.s.)}$, for various assumptions concerning the pseudoscalar form factor. We also obtain cross sections for these assumptions averaged over the LAMPF muon neutrino spectrum, and a differential cross section for this reaction.

Before beginning the actual calculation we should discuss a few differences between the electron neutrino reaction and the muon neutrino reaction to the $^{12}N_{(g.s.)}$ state. The matrix elements describing both reactions are, of course, very similar. However, because the mass of the electron is so small and because in $|M|^2$ all terms containing the pseudoscalar form factor are proportional to the lepton mass squared, all pseudoscalar terms are effectively suppressed. This is not true for the muon neutrino reaction due, of course, to the mass of the muon and the contribution of the pseudoscalar form factor cannot be ignored particularly in the energy region near threshold for this reaction. This will be apparent in our results and is the major difference between the matrix elements for the muon neutrino and electron neutrino reactions.

Finally, although the neutrino energy for the muon neutrino reaction is substantially higher than that for the electron neutrino reaction in present experiments, in the former case a muon must be produced. This effectively subtracts 105 MeV from the total available energy for the muon neutrino reaction whereas in the electron neutrino case only half an MeV is subtracted. Because the LAMPF muon neutrino spectrum falls rapidly with increasing neutrino energy, there is substantial overlap kinematically between the electron neutrino energy range given by the Michel spectrum and the LAMPF spectrum for in flight muon neutrinos.

We begin by considering the first order weak transition matrix element:

$$< \mu^- {}^{12}N_{(g.s.)}|H_w|\nu_\mu {}^{12}C > = \frac{G}{\sqrt{2}} \cos\theta_C \overline{u}_\mu \gamma^\lambda (1-\gamma_5) u_\nu < {}^{12}N_{(g.s.)}|J_\lambda(0)|{}^{12}C > .$$
(11)

which is entirely sufficient for this calculation.

The quantity $< {}^{12}N_{(g.s.)}|J_\nu(0)|{}^{12}C >$, may be written in terms for the vector and axial vector current matrix elements as:

$$< {}^{12}N_{(g.s.)}|J_\nu(0)|{}^{12}C > = < {}^{12}N_{(g.s.)}|V_\nu(0)|{}^{12}C > - < {}^{12}N_{(g.s.)}|A_\nu(0)|{}^{12}C >.$$
(12)

We make use of an elementary particle model treatment to write the matrix elements $< {}^{12}N_{(g.s.)}|V_\nu(0)|{}^{12}C >$ and $< {}^{12}N_{(g.s.)}|A_\nu(0)|{}^{12}C >$ in terms of four vectors with coefficients which are Lorentz scalars and are called form factors. This treatment[11,12] has appeared many times in the literature and we write the matrix elements as:

$$< {}^{12}N_{(g.s.)}|V_\nu(0)|{}^{12}C > = -i\sqrt{2}m_i \epsilon_{\nu\mu\rho\sigma} q^\mu \xi^\rho Q^\sigma \frac{F_M(q^2)}{2m_i 2m_p}$$
(13a)

and

$$< {}^{12}N_{(g.s.)}|A_\nu(0)|{}^{12}C > = \sqrt{2}m_i \left(\xi_\nu F_A(q^2) + q_\nu \xi \cdot q \frac{F_P(q^2)}{m_\pi^2} - Q_\nu \xi \cdot q \frac{F_E(q^2)}{2m_i m_p} \right).$$
(13b)

Here the quantity ξ_ν is the $^{12}N_{(g.s.)}$ polarization vector,

$$q_\nu = p_{f\nu} - p_{i\nu}, \qquad (14a)$$

$$Q_\nu = p_{f\nu} + p_{i\nu}, \qquad (14b)$$

m_i is the ^{12}C mass, m_f is the $^{12}N_{(g.s.)}$ mass, m_p is the proton mass and F_M, F_A, F_P, and F_E are the weak form factors. These are functions of q^2 and the effects of the nuclear structure are included in them. Clearly therefore the physics of this reaction is contained in the form factors and finding them accurately is absolutely necessary. However sufficient electron scattering and photo de-excitation data exists for the $^{12}C \leftrightarrow {}^{12}C^*$ (15.11 MeV) transition to determine the F_M form factor and this form factor is central to obtaining the others. This form factor[4] for $|q^2| \leq 3.7m_\pi^2$ and[13] for $3.7m_\pi^2 \leq |q^2| \leq 20m_\pi^2$ is obtained via the conserved vector current hypothesis and is given by:

$$F_M(q^2) = \frac{4.04 \cos^2\left(\frac{-q^2}{3.12m_\pi^2}\right)}{\left(1 - \frac{q^2}{2.86m_\pi^2}\right)^2} \qquad (15a)$$

for $|q^2| \leq 3.7m_\pi^2$ and

$$F_M(q^2) = \frac{252 \sin^2\left(\frac{(-q^2 - 3.7m_\pi^2)}{11.1m_\pi^2}\right)}{\left(1 + \left(\frac{q^4}{5m_\pi^4}\right)\right)^2} \qquad (15b)$$

for $3.7m_\pi^2 \leq |q^2| \leq 20m_\pi^2$. The data[13] used for these form factors is the most recent available and shows the diffractive minima better than earlier data. Equations (15a) and (15b) look quite different from the dipole fit, $\frac{F_M(0)}{\left(1 - \frac{q^2}{M^2}\right)^2}$, often assumed[14] but at low q^2, Eq.(15a) effectively becomes a dipole fit. This effectively determines the vector part of this weak process.

The form factors describing the axial current matrix element, Eq.(13b), are more difficult to determine. The axial current form factor, F_A is known accurately from beta decay of $^{12}N_{gs}$ and is given by:

$$F_A(0) = 1.03 \qquad (16)$$

and where the error is believed to be less than two percent. No direct measurements are available to provide a complete q^2 dependence for $F_A(q^2)$ although as we shall see muon-capture provides an additional constraint. However an argument due to Kim and Primakoff[15] originally based on a nucleons only impulse approximation but extended to include some pion-exchange current corrections yields:

$$\frac{F_A(q^2)}{F_A(0)} \simeq \frac{F_M(q^2)}{F_M(0)}. \qquad (17)$$

This formulation of $F_A(q^2)$ is in accord with muon capture results[4] which are very sensitive to this form factor which is responsible for over 80 percent of the muon capture rate and we shall use it here.

The form factor $F_P(q^2)$ has proved difficult to measure but arguments based on the partially conserved vector current hypothesis, PCAC, indicate that the form factor may be written as:

$$F_P(q^2) = \frac{m_\pi^2 F_A(q^2)}{m_\pi^2 - q^2}[1 + \xi(q^2)]. \qquad (18)$$

39

Some experimental evidence[16,17] from polarized muon capture experiments in ^{12}C as well as some theoretical work[15] is available to put limits on ξ. The errors are unfortunately large (both for experiment and theory) but a range of ξ from -1 to +1 more than covers present estimates[4].

Finally the form factor,$F_E(q^2)$ must be estimated. Arguments by Primakoff and Hwang[18] based on an impulse approximation treatment and similar to the arguments leading to Eq.(7) indicate that:

$$F_E(0) \simeq 3.75 \qquad (19a)$$

and

$$\frac{F_E(q^2)}{F_E(0)} \simeq \frac{F_M(q^2)}{F_M(0)}. \qquad (19b)$$

By direct calculation the contributions of the terms containing $F_E(q^2)$ as given by Eqs. (19a) and (19b) are at the one or two percent level and so we shall assume that it is given above to sufficient accuracy.

As we have noted earlier another important piece of data which involves the weak form factors described above is the muon capture reaction,$\mu^- + {}^{12}C \longrightarrow \nu_\mu + {}^{12}B_{gs}$. The $|q^2|$ value at which this muon capture reaction takes place is $q^2 = -.74\ m_\mu^2$. This q^2 value, which in terms of MeV may be written as $q^2 = -8260\ MeV^2/c^2$ would correspond to an incident neutrino energy of slightly above 200 MeV and an outgoing lepton angle of 20 degrees which is not unreasonable. Thus the q^2 value associated with the muon capture reaction would be a very appropriate one for determining the weak nuclear form factors for conditions relevant for the reaction considered here. Unfortunately experimental results[19-22] vary from 5700 sec^{-1} to 6300 sec^{-1}. Although this is not an extreme variation, it masks F_P. Experiments are underway[23] on the reaction $\mu^- + {}^3He \longrightarrow {}^3H + \nu_\mu$ are hoped to eventually yield results in the two to three percent range and it would be desirable to see the error for muon capture in ^{12}C similarly reduced.

The form factors given above yield muon capture results falling within the above experimental range. We believe that these form factors are reasonably accurate for $|q^2| < m_\pi^2$. They work well for muon capture for this and many similar nuclear transitions. However we suspect that equations (17) and (18) and (19b) break down at high $|q^2|$. The present LAMPF experiment has very little spectrum in the range where this might happen but it would be useful to have experimental results at larger $|q^2|$ to provide guidance.

Now that the form factors are determined we may calculate $|M|^2$,the transition matrix element squared. This is done by standard means and the result is:

$$M^2 = \frac{F_M^2}{8m_i^2 m_p^2} \left[Q^2[(\nu \cdot \mu)^2 - m_\mu^2(\nu \cdot \mu)] + (Q \cdot \nu)^2(\nu \cdot \mu - m_\mu^2) + (Q \cdot \mu)^2(\nu \cdot \mu) \right]$$

$$+ F_A^2 \left(\nu \cdot \mu + \frac{2(p_f \cdot \nu)(p_f \cdot \mu)}{m_f^2} \right)$$

$$+ \frac{F_P^2 m_\mu^2}{m_\pi^4}(\nu \cdot \mu) \left(2\nu \cdot \mu - m_\mu^2 + \frac{(p_f \cdot \nu)^2 + (p_f \cdot \mu)^2 - 2(p_f \cdot \nu)(p_f \cdot \mu)}{m_f^2} \right)$$

$$+ \frac{F_E^2}{16m_p^2 m_i^2}[2(Q \cdot \nu)(Q \cdot \mu) - Q^2(\nu \cdot \mu)]$$

$$\times \left(2\nu \cdot \mu - m_\mu^2 + \frac{(p_f \cdot \nu)^2 + (p_f \cdot \mu)^2 - 2(p_f \cdot \nu)(p_f \cdot \mu)}{m_f^2} \right)$$

$$+ \frac{F_M F_A}{m_i m_p}[(\nu \cdot \mu - m_\mu^2)(Q \cdot \nu) + (\nu \cdot \mu)(Q \cdot \mu)]$$

$$+ \frac{F_A F_P(2m_\mu^2)}{m_\pi^2} \left[-(\mu \cdot \nu) + \frac{(p_f \cdot \nu)(p_f \cdot \mu) - (p_f \cdot \nu)^2}{m_f^2} \right]$$

$$- \frac{F_A F_E}{2m_i m_p} \left[m_\mu^2(Q \cdot \nu) + \frac{p_f \cdot q}{m_f^2}[(Q \cdot \mu)(p_f \cdot \nu) + (Q \cdot \nu)(p_f \cdot \mu) - (\mu \cdot \nu)(Q \cdot p_f)] \right]$$

$$+ \frac{F_E F_P}{2m_\pi^2 m_i m_p} m_\mu^2(Q \cdot \nu) \left[2\nu \cdot \mu - m_\mu^2 + \frac{(p_f \cdot \nu)^2 + (p_f \cdot \mu)^2 - 2(p_f \cdot \mu)(p_f \cdot \nu)}{m_f^2} \right]$$

$$(20)$$

From this the cross section is immediately calculable and results may be obtained. We still however must take into account the final state electromagnetic interaction between the muon and the nucleus. We do this following a treatment due to Tzara[23] whereby this interaction is incorporated in a correction factor:

$$C_F \simeq \frac{2\pi \alpha Z m_\mu}{p_\mu} (1 - e^{\frac{-2\pi \alpha Z m_\mu}{p_\mu}}) \qquad (21)$$

where p_μ is the magnitude of the outgoing muon three-momentum. This treatment has yielded satisfactory results previously[4,11] and we include it as a factor in our cross section here.

In figure 1 we plot the cross section for the reaction $\nu_\mu + {}^{12}C \longrightarrow \mu^- + {}^{12}N_{(g.s.)}$ from threshold to 300 MeV. We do this for several different assumptions concerning the value of ξ in Eq.(8). We have already noted that polarized muon capture experiments place some limits on the value of ξ. These measurements for ξ carry errors of the order of 40 percent and yield results[16,17] for ξ of the order of 0 to 1. On the other hand theoretical estimates by Kim and Primakoff[15] placed ξ at around -.5 to -.3. In order to give some idea of the range of results which might be expected experimentally we have plotted the cross sections for $\xi = -1$, $\xi = -.5, \xi = 0$, and $\xi = 1$. The Los Alamos muon neutrino spectrum for the process described here has been previously published[10]. Averaging the neutrino cross section over this spectrum for the four assumptions for ξ used here yields:

$$< \sigma > = .694 \times 10^{-40} \ cm^2 \qquad (22a)$$

for $\xi = -1$,

$$< \sigma > = .643 \times 10^{-40} \ cm^2 \qquad (22b)$$

41

for $\xi = -.5$,

$$<\sigma> = .600 \times 10^{-40} \; cm^2 \qquad (22c)$$

for $\xi = 0$, and

$$<\sigma> = .537 \times 10^{-40} \; cm^2 \qquad (22d)$$

for $\xi = +1$. Here Eq.(12a) for $\xi = -1$ provides a baseline for no pseudoscalar interaction at all. Experimental results do indicate a pseudoscalar interaction and the value,$\xi = -.5$ represents a more realistic lower limit for ξ.

Because the uncertainty in the Los Alamos result is expected to be of the order of 30 percent, that experiment will not resolve the magnitude of the pseudoscalar contribution at present. However future experiments particularly if combined with better measurements for the reaction, $\nu_e + {}^{12}C \longrightarrow e^- + {}^{12}N_{(g.s.)}$, might help resolve this question. This is because the electron neutrino reaction is essentially independent of $F_P(q^2)$, as can be seen from Eq.(20) where all terms containing $F_P(q^2)$ are proportional to the lepton mass squared. Thus a precision measurement of $\nu_e + {}^{12}C \longrightarrow e^- + {}^{12}N_{(g.s.)}$ is essentially a measurement[4] of F_A, particularly at lower energy. This would enable the error in this calculation to be reduced and would therefore make conclusions concerning the pseudoscalar contribution firmer. We note that the calculation of reference 14 corresponds to a ξ of roughly -.5, which would fall in the range suggested by reference 17.

A discussion of errors should accompany these numbers. The axial current form factor dominates the cross section,providing over 80 percent of its total value. The primary ingredients which determine $F_A(q^2)$, are $F_A(0)$ and the q^2 dependence of F_A as given by Eq.(16) and Eq.(17) respectively. An analysis of muon capture results[24] indicated that the model given by Eq.(17) produced muon capture rates accurate to 10 to 15 percent. The error in the other form factors provides part of this error. We believe that our present calculation which makes use of essentially the same matrix element should have a similar error associated with it. If we take a very conservative position,because $|q^2|$ is larger over part of our energy range than that for muon capture, the errors may be be somewhat greater than those given by the analysis of reference 27. We therefore are reasonably certain that the above cross sections are accurate to the 15 to 20 percent range. Other reasonable considerations may lower this error however. In arriving at the 15 to 20 percent level of accuracy we have conservatively assumed the largest errors for all components of our calculation. However, as we shall see, the cross section,$\sigma(E_\nu)$, is dominated over the range of neutrino energy considered here by the low $|q^2|$ part of the form factors, or from Eq.(7) by the low $|q^2|$ part of $F_M(q^2)$ given by Eq.(5a). Because the errors on F_M are relatively small in this q^2 range, the actual error in the cross section calculation may be closer to the 10 percent level.

From the above discussion, the cross section is not sensitive to the values of the form factors at higher values of $|q^2|$. The situation would be very different for a differential cross section measurement. In figure 2 we show the differential cross section for the reaction,$\nu_\mu + {}^{12}C \longrightarrow \mu^- + {}^{12}N_{(g.s.)}$ for an incident muon neutrino energy of 200 MeV with ξ set equal to zero. As can be seen there is a striking minima in the region of θ_μ varying from 85 to 105 degrees as would be expected from Eqs.(5a) and 5(b) and which would not be present in a dipole type fit. The presence of such a dip would be important evidence for whole nucleus CVC at higher $|q^2|$. Although an experiment to measure a differential cross would be very difficult and would

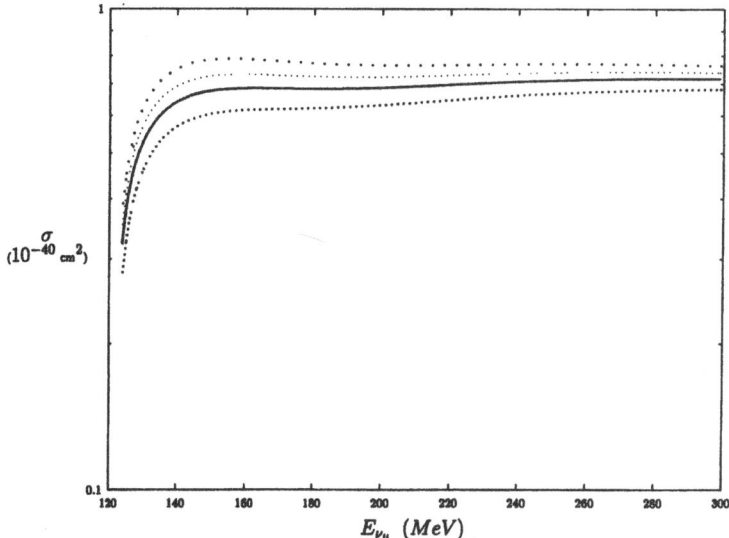

Fig. 1 Plot of the cross section, σ, as a function of neutrino energy for the reaction,$\nu_\mu +$ $^{12}C \longrightarrow \mu^- + {}^{12}N_{(g.s.)}$. The dense dotted curve indicates a value used for ξ of 1. The solid curve indicates a value used for ξ of 0. The light dotted curve indicates a value used for ξ of -.5. The thinly dotted curve indicates a value used for ξ of -1. The parameter ξ is defined in Eq.(18).

necessarily have to run over a very long time, it is not intrinsically impossible. We remark however that the present Los Alamos experiment will not resolve this question. In figure 2 we have also plotted the differential cross section averaged over Los Alamos muon-neutrino spectrum. As can be seen the dip is washed out by the average. A spectrum sharply peaked above the threshold region will be needed to observe this dip. The Los Alamos group has been obtaining data for the transition,$\nu_\mu + {}^{12}C \longrightarrow$ $\mu^- + {}^{12}N_{(g.s.)}$ and now have about 60 events. Preliminary indications favor a result for the cross section in the range of .6 $\times$ 10^{-42} cm^2 which is well in accord with the above range. Because agreement between theoretical calculations and experimental results for the electron neutrino reaction $\nu_e + {}^{12}C \longrightarrow e^- + {}^{12}N_{(g.s.)}$ is quite good[28], and because the theoretical calculations for the muon neutrino and electron neutrino reactions make use of the same assumptions, it would be surprising if theory and experiment diverged substantially for the muon neutrino case. Such a divergence would raise questions concerning the inclusive reaction,$\nu_\mu + {}^{12}C \longrightarrow X + \mu^-$.

It might have been hoped that a measurement of the cross section for $\nu_\mu + {}^{12}C \longrightarrow$ $\mu^- + {}^{12}N_{(g.s.)}$ would settle the question of the accuracy of the neutrino spectrum used for the Los Alamos Experiment. This is unfortunately not true. The spectrum averaged cross section is given by:

$$< \sigma >= \frac{\int_{threshhold}^{E_\nu^{max}} \sigma(E_\nu)\phi(E_\nu)\,dE_\nu}{\int_{threshold}^{E_\nu^{max}} \phi(E_\nu)\,dE_\nu} \tag{23}$$

where E_ν^{max} is the maximal neutrino energy,approximately 300 MeV here, and $\phi(E_\nu)$ is the muon neutrino spectrum. However,except in the region near threshold, as can

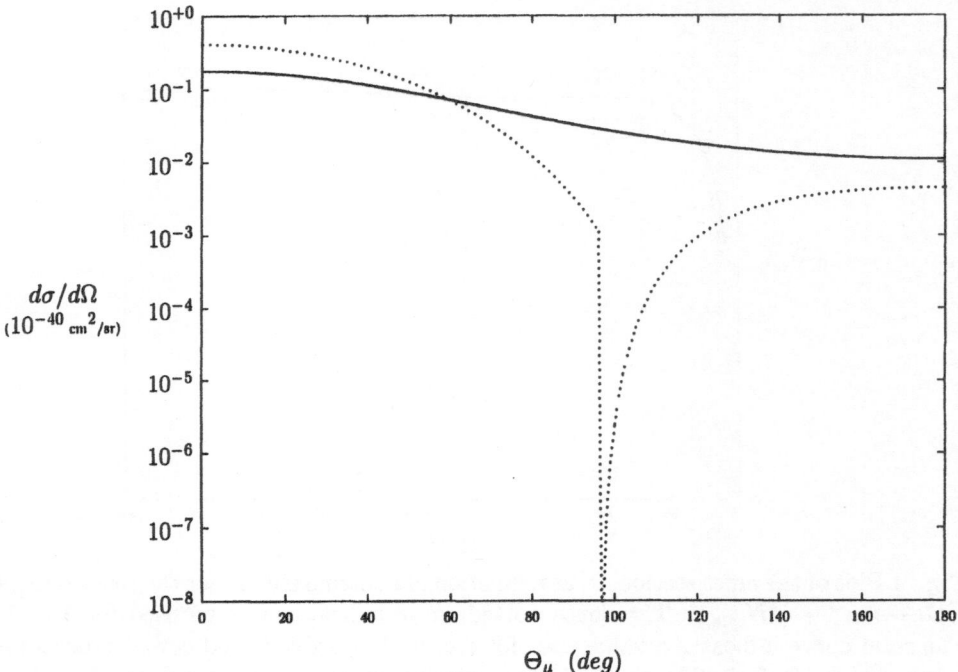

Fig. 2 Plot of the differential cross section for the reaction $,\nu_\mu + {}^{12}C \longrightarrow \mu^- + {}^{12}N_{(g.s.)}$, as a function of outgoing muon laboratory angle. The dotted curve indicates the unaveraged differential cross section for a neutrino of energy of 200 MeV with $\xi = 0$. The solid curve indicates the differential cross section averaged over the Los Alamos decay-in-flight muon neutrino spectrum again with $\xi = 0$.

be seen from figure 1, the cross section is quite flat. Thus we find:

$$< \sigma > \approx \frac{\sigma \int_{threshold}^{E_\nu^{max}} \phi(E_\nu)\, dE_\nu}{\int_{threshold}^{E_\nu^{max}} \phi(E_\nu)\, dE_\nu} \tag{24}$$

or

$$< \sigma > \approx \sigma. \tag{25}$$

Thus except in the region about threshold, where the cross section is smallest there is little sensitivity to the shape of the spectrum. In fact it would be a convincing test of the calculation presented here if the experimental results are always the same for the averaged cross section irrespective of the spectrum used and always in the neighborhood of $.6 \times 10^{-40}\ cm^2$. It is to be regretted that this is so because a few of the theoretical treatments for the reaction, $\nu_\mu + {}^{12}C \longrightarrow X + \mu^-$, were sensitive to variations in the spectrum, particularly at higher energies.

We therefore expect the averaged cross sections, $< \sigma >$ to be in the range of $0.6 \times 10^{-40}\ cm^2 \pm 20\ \%$. The experimental result which is in this range is very encouraging not only from the view point of theoretical efforts but also for the enhanced credibility of related experiments concerning the inclusive reaction in ${}^{12}C$. Unfortunately this process does not provide reliable additional knowledge concerning the muon neutrino spectrum used by LAMPF. However other neutrino reactions in nuclei might serve as a more useful test of the spectrum. Among these are reactions in the ${}^3H \leftrightarrow {}^3He$ system which we shall now discuss.

44

THE $^3H \leftrightarrow ^3He$ SYSTEM

Neutrino reactions on ^3H and anti-neutrino reactions on ^3He might be interesting for a number of reasons. First in either case these transitions are mirror transitions. Thus the initial and final state overlaps are large. Secondly in the former case the percentage of neutrons on which neutrinos might interact is greater for ^3H than for any other light nucleus on which experiments might actually be performed. A similar statement concerning protons and anti-neutrinos can be made for the latter case. Finally it will turn out that the some of the interference terms are quite large and this will strongly favor the neutrino reaction over the anti-neutrino one. Nonetheless the anti-neutrino reaction is quite large when compared to the ^{12}C case. We now briefly discuss these reactions. In what follows we shall handle the calculations for both simultaneously. We first make a few remarks concerning anti-neutrino reactions.

Anti-neutrino reactions in nuclei have received less attention in general than neutrino reactions in nuclei. This is largely because of the current effort to understand the solar neutrino deficit but is also due to the greater difficulty in obtaining anti-electron neutrino spectra at facilities[1,2] such as LAMPF and KARMEN. Because both groups make use of proton accelerators to produce pions, the number of π^+ produced will be greater than the number of π^-. In addition the Michel spectra used by the LAMPF and KARMEN groups are produced by π^+ which decay at rest to μ^+ (and a muon neutrino) which then decay to electron neutrinos, anti-muon neutrinos, and, of course, positrons. Typically the π^- are captured in the beam dump[25] before producing μ^-. The beam dump at LAMPF consists for example largely of Fe and Cu. In addition any μ^- produced is liable to be captured thus producing muon neutrinos but no electron anti-neutrinos.

Nonetheless, there are strong reasons for studying the anti-neutrino reaction in nuclei. Reactors provide an excellent source of low energy electron anti-neutrinos and at least one group is considering the possiblilty of using reactor anti-neutrinos as part of their calibration efforts for a $^7Li - ^{127}I$ solar neutrino detector to gain experience in low energy neutrino reactions. In addition it is crucially important to accurately determine the anti-electron neutrino background which might exist in current and planned oscillation experiments at LAMPF and KARMEN. The assumptions made concerning the capture of π^- and μ^- particles are largely estimates and any error is them might have consequences for the oscillation measurements.

Furthermore, there has been some interest[26] in the construction of an ^3He detector in order to observe breakup neutrino reactions. This type of detector would be particularly useful for observing pp solar neutrinos. Both liquid and gas ^3He detectors have been proposed and a prototype is under development. The anti-neutrino background from terrestrial and atmospheric sources would be of some use to this project.

Another problem involved in studying the $^3H \leftrightarrow ^3He$ system is on the neutrino reaction side of the picture. The nucleus, ^3H, is, of course, not stable. However it does have a long half life of the order of 12 years and has been used for a target for electron scattering. Although electron scattering experiments take much less time than neutrino experiments, we shall see that the cross sections are sufficiently large that the attempt might be worth making.

Although the weak transition, $^3He \leftrightarrow ^3H$, and the isotopically related transitions $^3He \leftrightarrow ^3He$, and $^3H \leftrightarrow ^3H$ have been studied extensively[27,28,29] and the three nucleon system is,in principle, a simple one, a number of questions still remain concerning it. In a recent paper[28], Fearing and Congleton noted that there was a discrepancy between standard shell model impulse approximation calculations for the muon capture reaction $\mu^- + {}^3He \longrightarrow \nu_\mu + {}^3H$ and an elementary particle model calculation for the same reaction. They noted that the elementary particle model gave a very accurate fit to present measurements but that the impulse approximation calculation was about 15 percent below the expected rate and indicates work remains to be done for the understanding of even a small nucleus such as ${}^3He$ or ${}^3H$.

Just as for the ${}^{12}C$ case described previously, the processes we are considering can be well described as a first order in the weak interaction. We are going to be mainly interested in incoming neutrinos with energies up to 1 GeV so that values of q^2 of the order of only a few GeV are possible. This is well below the intermediate vector boson mass squared so that the boson propagator behaves as a constant. We therefore write the first order weak interaction in its standard form as:

$$< \mu^- \; {}^3He|H_w|\nu_\mu \; {}^3H > = \frac{G}{\sqrt{2}} \cos\theta_C \bar{u}_\mu \gamma^\lambda (1 - \gamma_5) u_\nu <^3 He|J_\lambda(0)|^3H > . \quad (26)$$

We have written the above matrix element in terms of the muon and the muon neutrino but we can write this for any charged lepton and corresponding lepton neutrino by simply substituting for the muon quantities. In the above equation, the lepton part is entirely known but the hadronic matrix element, $<^3 He|J_\lambda(0)|^3H >$, must be determined.

The hadronic current current, $J_\mu(0)$ is written as:

$$J_\mu(0) = V_\mu(0) - A_\mu(0) \quad (27)$$

where V_μ and A_μ are the vector and axial vector parts of the weak nuclear current. Thus if we knew the matrix elements of these currents we could immediately calculate our cross sections. Several methods of calculation are available. Here we make use of an elementary particle model which has given particularly accurate results[28] as was noted above in describing the muon capture reaction $\mu^- + {}^3He \longrightarrow \nu_\mu + {}^3H$. We have previously[27] used this model in calculating the electroweak process, $e^- + {}^3He \longrightarrow \nu_e + {}^3H$ for which it produced reasonable results. In this model the matrix elements of the nuclear vector and axial vector currents are written in a Lorentz invariant way in terms of spinors for the spin one-half nuclei ${}^3He$ and ${}^3H$, linearly independent vectors, and form factors. These matrix elements are well known[27] and may be written as:

$$<^3 He|V_\mu(0)|^3H > = \bar{u}_f[\gamma_\mu F_V(q^2) + i\frac{F_M(q^2)\sigma_{\mu\nu}q^\nu}{2m_p}]u_i \quad (28a)$$

and

$$<^3 He|A_\mu(0)|^3H > = \bar{u}_f[\gamma_\mu\gamma_5 F_A(q^2) + \frac{q_\mu\gamma_5 F_P(q^2)}{m_\pi}]u_i \quad (28b)$$

where i is the initial nucleus and f is the final nucleus, ${}^3He$ and ${}^3H$ respectively. The structure of the nucleus is contained in the four form factors, $F_V(q^2), F_M(q^2), F_A(q^2)$, and $F_P(q^2)$. Thus a determination these form factors would enable us to write and evaluate a transition matrix element for the ${}^3H \leftrightarrow {}^3He$ transition.

The form factors of the vector current matrix element may be evaluated by making use of the conserved vector current hypothesis. We have already done this in reference 27 and we shall merely quote the results for $F_V(q^2)$ and $F_M(q^2)$, namely:

$$F_V(q^2) = F_V(0) \cos^2(\frac{-q^2}{17.1m_\pi^2})(1 - \frac{q^2}{6.25m_\pi^2})^{-2} \quad (29a)$$

for $|q^2| \leq 24.5 m_\pi^2$

$$F_V(q^2) = F_V(0) \cos^2\left(\frac{-q^2}{14.39 M_\pi^2}\right)\left(1 - \frac{q^2}{4.35 m_\pi^2}\right)^{-2} \tag{29b}$$

for $|q^2| > 24.5 m_\pi^2$ and

$$F_M(q^2) = F_M(0) \cos^2\left(\frac{-q^2}{28.4 M_\pi^2}\right)\left(1 - \frac{q^2}{4.5 m_{p^i}^2}\right)^{-2} \tag{30a}$$

for $|q^2| \leq 43.0 m_\pi^2$

$$F_M(q^2) = F_M(0) \cos^2\left(\frac{-q^2}{26.0 m_\pi^2}\right)\left(1 - \frac{q^2}{3.5 m_\pi^2}\right)^{-2} \tag{30b}$$

for $|q^2| > 43.0 m_\pi^2$.

We note that at low values of $|q^2|$, the cosine terms in eqs. (29a),(29b),(30a), and (30b) are approximately one, yielding the dipole form factors familiar from earlier work[29,30]. At higher $|q^2|$, the cosine terms model the diffractive minimum noted in the electromagnetic form factors. We shall discuss this point later.

The axial current form factors are unfortunately not as well known as the vector current form factors. Of the two axial current form factors, $F_A(q^2)$ and $F_P(q^2)$, the former makes a large contribution to the cross section as will be seen. It should be noted however that all terms in the transition matrix element squared which contain the factor $F_P(q^2)$ also contain the charged lepton mass squared as a factor. For the reaction, $\nu_e + {}^3H \longrightarrow e^- + {}^3He$ where the charged lepton is the electron, terms proportional to F_P are suppressed and do not contribute in a measureable way to the transition matrix squared and we may ignore them. But for the reactions, $\nu_\mu + {}^3H \longrightarrow \mu^- + {}^3He$, and $\bar\nu_\mu + {}^3He \longrightarrow \mu^+ + {}^3H$, for which the charged lepton is a muon, the $F_P(q^2)$ terms may not be ignored a priori.

We must determine $F_A(q^2)$ which, as noted, does contribute substantially to the transition matrix squared. The quantity $F_A(0)$ can be determined[28] from the beta decay, ${}^3H \longrightarrow {}^3He + e^- + \bar\nu$, which takes place at $q^2 \simeq 0$. There is no direct way to determine the q^2 dependence of F_A. However, as previously mentioned, a result by Kim and Primakoff[30] based upon the impulse approximation leads to Eq.(17) which we use here. Kim and Primakoff extended the original calulation[15] to include some of the exchange current contributions.

The use of Eq.(17) is reasonable for the processes considered here based upon the derivation of this result. However, for all derived results it is necessary to test the utility of the result by comparison with actual experiment. We find that this relations works extremely well for muon capture[28,29] in ${}^3He$ as well as for muon capture[15,31] in ${}^6Li$ and in ${}^{12}C$, producing results as good or better than other currently used methods. Because the form factor, F_A, dominates[32] the muon capture result for all of the processes mentioned above, our confidence in Eq.(17) is increased.

An expression for $F_P(q^2)$ is still needed, and to obtain it we make use of the partially conserved vector current hypothesis to write:

$$F_P(q^2) = \frac{-m_\pi(M_i + M_f)F_A(q^2)}{(q^2 - m_\pi^2)}(1 + \xi(q^2)). \tag{31}$$

This result is well known[15] but is difficult to fully confirm due to the fact that muon capture and neutrino reactions are not very sensitive to $F_P(q^2)$. Theoretical arguments[15] indicate that for the $^3He \leftrightarrow ^3H$ transition, ξ is small and so we set it to zero in Eq.(31). We may now proceed with our calculation. We note that the values for the necessary form factors at $q^2 = 0$ have been previously determined[27] and are given by:

$$F_V(0) = 1.0 \qquad (32a)$$

$$F_M(0) = -5.44 \pm .0015 \qquad (32b)$$

and

$$F_A(0) = -1.212 \pm .004 \qquad (32c)$$

where eqs. (32a) and (32b) are determined by CVC from electron scattering data and eq.(32c) is determined from beta decay data as has already been noted. We remark that the errors associated with the form factors at $q^2 = 0$ are quite small and in the case of $F_V(0)$ would be zero because it is just the difference in charge between the initial and final state nucleus. At larger $|q^2|$ the errors are mostly due to an uncertainty in the mass in the dipole part of the form factors. This gives gives rise to uncertainies of the order of 8 to 9 percent at worst at $q^2 \approx -m_\mu^2$. These uncertainties are likely to increase at higher $|q^2|$.

We may obtain the matrix element squared for the processes under consideration here via eq.(26) and its analogue for the anti-neutrino processes such as, $\bar{\nu}_e + {}^3He \longrightarrow e^+ + {}^3H$ similarly. The result is:

$$
\begin{aligned}
|M|^2 = \frac{1}{m_\mu m_\nu} \Bigg[&\frac{4|F_V|^2}{M_i M_f} [p_f \cdot \nu p_i \cdot \mu + p_f \cdot \mu p_i \cdot \nu - \mu \cdot \nu M_i M_f] + \\
&\frac{4|F_A|^2}{M_f M_i} [p_f \cdot \nu p_i \cdot \mu + p_f \cdot \mu p_i \cdot \nu + \mu \cdot \nu M_i M_f] + \\
&\frac{2|F_M|^2}{m_p^2 M_i M_f} [\mu \cdot \nu (p_i \cdot \mu p_f \cdot \mu + p_i \cdot \nu p_f \cdot \nu + \mu \cdot \nu M_i M_f) + \\
&(-p_i \cdot \nu p_f \cdot \nu - \frac{1}{2}\mu \cdot \nu - \frac{3}{2}\mu \cdot \nu M_i M_f) m_\mu^2] + \\
&\frac{2|F_P|^2}{M_i M_f m_\pi^2} [m_\mu^2 \mu \cdot \nu (p_i \cdot \nu - p_i \cdot \mu - M_i(M_f - M_i))] - \\
&\frac{8 F_V F_A}{M_i M_f} [(p_i \cdot \nu + p_i \cdot \mu)\nu \cdot \mu - m_\mu^2 p_i \cdot \nu] + \\
&\frac{4 F_M F_V}{m_p M_f M_i} [\mu \cdot \nu (p_i \cdot \nu - p_i \cdot \mu)(M_f - M_i) + 2(\mu \cdot \nu)^2 + \\
&\qquad - m_\mu^2 (\frac{1}{2} p_i \cdot \nu (M_f - M_i) + \frac{3}{2} M_i \mu \cdot \nu] + \\
&\frac{-4 F_A F_M}{m_p M_f M_i} [(\mu \cdot \nu (p_f \cdot \mu + p_f \cdot \nu) M_i + \\
&\qquad \mu \cdot \nu (p_i \cdot \mu + p_i \cdot \nu) M_f \\
&\qquad - m_\mu^2 (p_f \cdot \nu M_i + p_i \cdot \nu M_f)] + \\
&\frac{-4 F_A F_P}{M_f M_i m_\pi} [m_\mu^2 (p_i \cdot \nu (M_f - M_i) + M_i \mu \cdot \nu)] \Bigg]
\end{aligned}
\qquad (33)
$$

In the above equation, μ_α is the muon four-momentum, M_f and M_i are the final and initial state nuclear masses respectively and ν_α is the neutrino four momentum. We have written this matrix element squared for the muon case but it will serve for all the charged leptons and their respective neutrinos by simply substituting for the muon mass. We note that the only difference between the neutrino case shown here and the anti-neutrino case is that the $F_V F_A$ and $F_M F_A$ change sign. This turns out to be important. We note that every term in this matrix element squared is proportional to the neutrino energy and in the limit in which the charged lepton mass may be ignored, they are also proportional to the charged lepton energy. The appropriate cross sections, and differential cross sections may now be calculated in the usual way.

We are now able to obtain the total cross sections for the reactions $\nu_e + {}^3H \longrightarrow e^- + {}^3He$, and $\nu_\mu + {}^3H \longrightarrow \mu^- + {}^3He$ and the anti-neutrino analogues, $\bar{\nu}_e + {}^3He \longrightarrow e^+ + {}^3H$ and $\bar{\nu}_\mu + {}^3He \longrightarrow \mu^+ + {}^3H$. In figure 3 we show the total cross section for the reactions, $\nu_\mu + {}^3H \longrightarrow \mu^- + {}^3He$, and $\nu_e + {}^3H \longrightarrow e^- + {}^3He$ from threshold to 1 GeV. We also display the for the muon neutrino case, the contributions to the cross sections from the terms in Eq.(33) proportional to F_A^2, F_V^2, F_M^2, and F_P^2. In figure 4 we show the the total cross section and also the contributions from the terms in Eq.(33) proportional to $F_V F_A, F_M F_A, F_V F_M$, and $F_A F_P$, i.e. all contributing interference terms for the muon neutrino case. In figures 5 and 6 we present the analogous results for the anti-neutrino case. Finally in figure 7 we obtain the differential cross section for this reaction for incoming neutrino energies of 1.5 GeV.

Because the reactions, $\nu_e + {}^3H \longrightarrow e^- + {}^3He$, and $\bar{\nu}_e + {}^3He \longrightarrow e^+ + {}^3H$, might be observed over a Michel spectrum, we calculate the flux averaged cross section for these reactions, given by:

$$< \sigma > = \frac{\int_{th}^{52.8} \sigma(E_\nu) \phi(E_\nu), dE_\nu}{\int_{th}^{52.8} \phi(E_\nu), dE_\nu} \tag{34}$$

where $\phi(E_\nu)$ is given by:

$$\phi(E_\nu) = N \frac{12 E_\nu^2}{(52.8)^4}(52.8 - E_\nu). \tag{35}$$

We note that the normalization of the flux is sometimes taken from zero to 52.8 MeV. This has caused some confusion in results presented for neutrino reactions in ${}^{12}C$ but in this case, threshold is sufficiently low that there is not much difference between the two normalizations. Using the normalizations given by eqs. (34) and (35) the flux average cross sections are:

$$< \sigma > = .927 \times 10^{-40} \; cm^2 \tag{36a}$$

for the neutrino case and

$$< \sigma > = .625 \times 10^{-40} \; cm^2 \tag{36b}$$

for the anti-neutrino case.

In looking at figures 3 and 5 one immediately observes that the cross sections are quite large. For an energy of approximately 150 MeV, the cross section for the reaction $\bar{\nu}_\mu + {}^3He \longrightarrow \mu^+ + {}^3H$, is approximately $3 \times 10^{-40} cm^2$, whereas that for the reaction $\nu_\mu + {}^{12}C \longrightarrow \mu^- + {}^{12}N_{(g.s.)}$ it is $.6 \times 10^{-40} \; cm^2$ i.e. the former reaction has a cross section approximately 5 times as large as that for ${}^{12}C$. For the neutrino case the result is even more striking. The neutrino reaction at the same energy is approximately twice as large as the anti-neutrino reaction and is an order of magnitude larger than the ${}^{12}C$ case. Similar results are true over the Michel spectrum. Again this is not surprising as this transition $^3He \leftrightarrow^3 H$, is a mirror transition whereas the $^{12}C \leftrightarrow ^{12}N_{(g.s.)}$ transition is not. Also from Eqs.(17) and (32c) it is clear that the magnitude of $F_A(q^2)$ is of the order of that for $F_V(q^2)$. This is an important point because for the mirror transition $^{13}C \leftrightarrow^{13} N_{gs}$, F_A is relatively small[32] which led to smaller cross sections than were initially expected. At low q^2, as is well known[12], the transition matrix element squared contains terms of the form:

$$3F_A^2 + F_V^2. \tag{37}$$

These terms are the leading ones in the transition matrix element and so it is clear that the axial current form factor dominates the low energy part of the neutrino cross section. This can clearly be seen in figures 3 and 5. Furthermore as q^2 increases,the relative contributions of F_A^2 and F_V^2 become equal (although $F_A(0)$ is of course larger than $F_V(0)$). This behavior is also visible in figures 3 and 5.

A perhaps unexpected feature of these reactions is the relatively large size of some of the interference terms. For both the neutrino and anti-neutrino reactions, the intereference term proportional to $F_A F_M$ is particularly large and positive for the neutrino reaction but negative for the anti-neutrino reaction. It is the size of this term which prevents the anti-neutrino cross section from becoming remarkably large. In addition the interference term proportional to $F_A F_P$ is relatively large at lower E_ν for the reactions, $\nu_\mu +^3 H \longrightarrow \mu^- +^3 He$ and $\bar{\nu}_\mu +^3 He \longrightarrow \mu^+ + ^3 H$. From threshold to around 120 MeV, this interference term is of the order 20 percent of the cross section. It falls with increasing incident neutrino energy. This may be seen in figure 4. Thus if neutrino experiments of high accuracy near threshold become possible, a measurement of F_P might become possible. This would be extremely useful as it is very difficult to measure the pseudoscalar form factor and questions still remain concerning its size. Because as was noted earlier,all terms in the transition matrix squared which contain F_P are also proportional to the charged lepton mass squared, the $F_A F_P$ terms are not measureable for the reactions, $\nu_e +^3 H \longrightarrow e^- +^3 He$ and $\bar{\nu}_e + ^3 He \longrightarrow e^+ + ^3 H$ as can be seen from figure 6.

A few other features of these reactions are worth noting. As has been observed in a number of cases[33], exclusive neutrino cross sections become flat at energies well above threshold. This is because the fall off in the form factors is matched by the increase in both incoming and outgoing lepton energies as the incident neutrino energy increases. This constancy away from threshold makes them unsuitable for measuring a spectrum away from threshold as has been noted (see Eqs.(23),(24),and (25)). However, because the masses in the dipole parts of $F_V(q^2), F_M(q^2)$, and $F_A(q^2)$ are much larger for the ^3He and ^3H cases than for the ^{12}C, of the order of 4 to 6 m_π^2 as opposed to 2 to 3 m_π^2 respectively,this causes the flat asymptotic behavior to be reached at a higher E_ν for these cases than for the ^{12}C case. Thus a neutrino reaction in ^3H or ^3He might be a useful test of a neutrino or anti-neutrino spectrum. The flatness of the cross sections with increasing neutrino energies are best seen in figure 5.

As was noted earlier we expect errors in the cross sections to be of the order of 8 to 9 percent, based mostly on the uncertainties in the form factors F_M and F_V obtained from electron scattering data. We should remark however that the form factors are best known at low q^2 and least well known in the region of the diffractive minimum. However, the largest contributions to the cross section come from the form factors at small q^2. This due to the fact that the form factors are largest at small q^2 as can be seen from Eqs.(29a),(29b),(30a), and (30b). However at small q^2 the energy and momentum of the outgoing lepton E_μ and $|\vec{p}_\mu|$, are largest. Because the

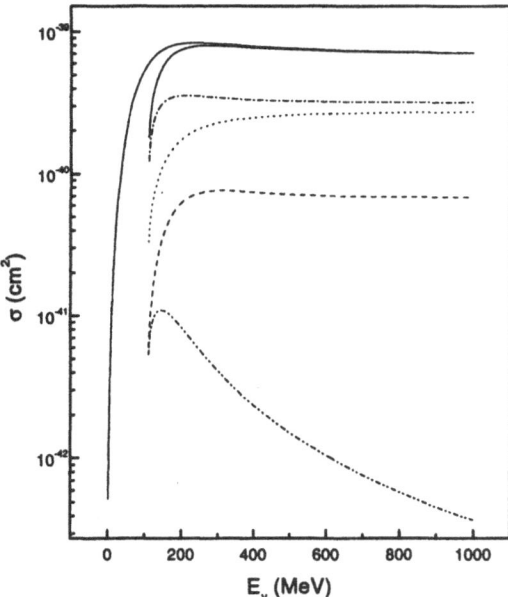

Fig. 3 Plot showing the reaction,$\nu_\mu +^3 H \longrightarrow \mu^- +^3 He$ from threshold to 1 GeV. The solid curve is the total cross section. The dashed dot curve is the contribution from the $F_A(q^2)^2$ term alone. The dotted curve is the contribution from the $F_V(q^2)^2$ term alone. The dashed curve is the contribution from the $F_M(q^2)^2$ term alone and the dashed double dotted curve is the contribution from the $F_P(q^2)^2$ term alone. The top solid curve is the total cross section for the reaction,$\nu_e +^3 H \longrightarrow e^- +^3 He$.

cross section,$\sigma(E_\nu)$, is primarily driven by the kinematic factor $E_\mu|\vec{p}_\mu|$, and the form factors, the largest contribution to the cross section comes from the region where the fit is essentially dipole and where the errors are smallest. Thus the the 8 to 9 perecent error mentioned earlier represents a worst case error.

We have noted above that the cross section $\sigma(E_\nu)$, is not sensitive the the behavior of the form factors at higher q^2. However a differential cross section might enable the form factors to be more directly viewed. In figure 6 we plot the differential cross section for an incident neutrino energy of 1.5 GeV. Here sharp dips from the diffractive minima in the form factors are clearly visible. Unfortunately, these minima are washed out in neutrinos produced in a broad spectrum such as has been available at LAMPF. A sharp spectrum at high energy might allow observation of such structure. The presence of such structure would be important evidence for whole nucleus CVC at high energy. However a differential neutrino cross section measurement would at present be extremely difficult.

Finally we should remark that although the $^3H \leftrightarrow^3 He$ system is a relatively simple one, it cannot be said to be fully understood. In connection with the closely related weak process $\mu^- +^3 He \longrightarrow^3 He + \nu_\mu$, Fearing and Congleton investigated different models for calculating the capture rate for this reaction[28]. In particular they used an impulse approximation shell model based calculation and an elementary particle model calculation. They found very good agreement between the elementary particle model calculation and the measured capture rate but they impulse approximation calculation produced results about 15 percent smaller than expected. Further analysis by them indicated that this difference came from missing magnetic strength in the impulse approximation model. Clearly it would be highly desirable to have

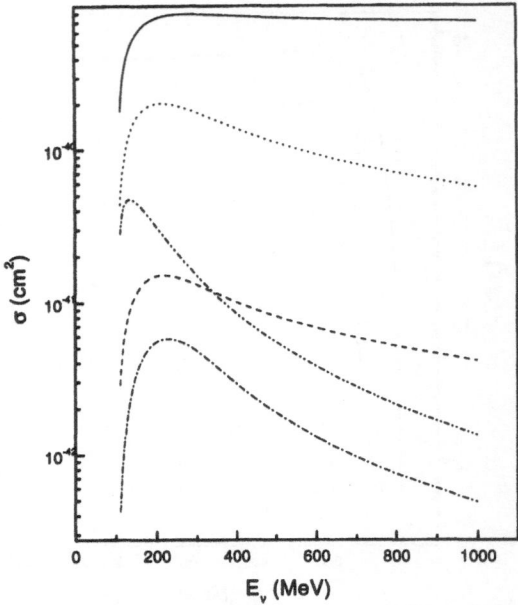

Fig. 4 Plot showing the cross section for the reaction, $\nu_\mu + ^3H \longrightarrow \mu^- + ^3He$ from threshold to 1 GeV. The solid curve is the total cross section. The dotted curve is the contribution from the $F_M(q^2)F_A(q^2)$ interference term . The dashed curve is the contribution from the $|F_V(q^2)F_A(q^2)|$ interference term. The dashed dotted curve is from the $|F_V(q^2)F_M(q^2)|$ interference term. The dashed double dotted curve is the contribution from the $|F_A(q^2)F_P(q^2)|$ interference term. All quantities in absolute value signs are negative.

additional experimental data to study this point. Although neutrino measurements are admittedly not easy there are not many other possibilities. The inverse beta decay $e^- + \, ^3He \longrightarrow \nu_e + \, ^3H$ would be exceedingly difficult[27,34] and parity violating electron scattering is not sensitive to the appropriate nuclear form factors[35]. Thus an anti-neutrino experiment on ^3He would be very useful and at least the cross section is relatively large for a weak nuclear exclusive process. Clearly much work remains to be done to provide a clear understanding of weak processes even in simple nuclei.

CONCLUSIONS

Thus we may conclude that we do not understand nuclear neutrino reactions as well as we would like. Our understanding of exclusive processes at least in light nuclei is reasonable (i.e. at the 20 % level) both experimentally and theoretically. Our understanding of inclusive processes is less good. Although the low energy inclusive process at Michel spectrum energies for ^{12}C seems to be reasonably well understood, the same cannot be said for higher energy muon neutrino spectra of the type formerly available at LAMPF. It would be enormously useful to have both a muon and electron neutrino beam available in the threshold to 500 MeV region. If this were available for both electron and muon neutrinos at 5 percent accuracy, a determination of the pseudoscalar contributions to the exclusive reaction cross section might be possible. It would also obviously help in understanding the inclusive reaction.

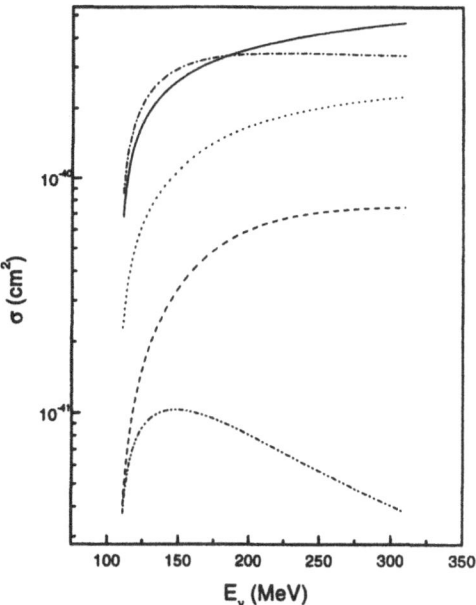

Fig. 5 Plot showing the cross section for the reaction, $\bar{\nu}_\mu +^3 He \longrightarrow \mu^+ +\ ^3H$, from threshold to 300 MeV. The solid line is the total cross section and the other lines show the contributions from the terms proportional to the form factors squared following the scheme of figure 3.

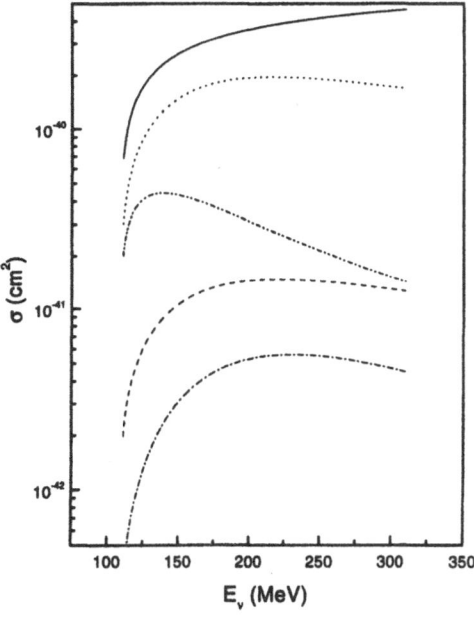

Fig. 6 Plot showing the cross section for the reaction $\bar{\nu}_\mu +^3 He \longrightarrow \mu^+ +\ ^3H$, from threshold to 300 MeV. The solid line is the total cross section. The other lines show the contributions of the intereference terms following the scheme of figure 4.

Finally we note that it would be very useful to have some anti-neutrino data. Although, except at reactor energies, anti-neutrino spectra are more difficult to obtain

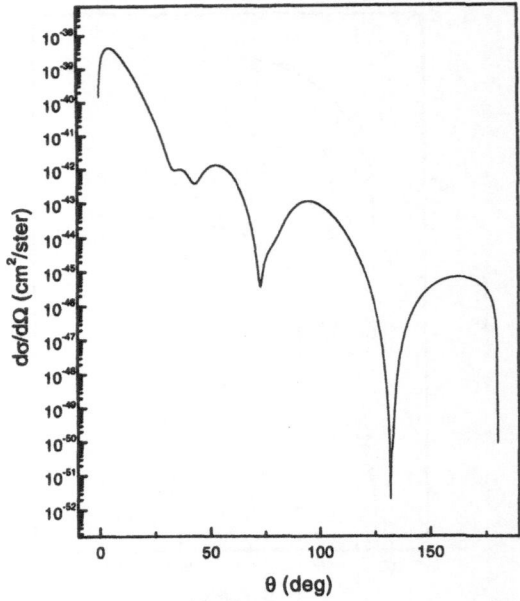

Fig. 7 Plot showing the differential cross section for the neutrino reaction for an incoming neutrino energy of 1.5 GeV.

than neutrino spectra they would provide a complimentary set of information to the present data. In particular for a mirror nucleus system such as the $^3H \leftrightarrow ^3He$ system described here, the difference between the neutrino and anti-neutrino cross section depends primarily on large interference terms which might be determined. This would be an excellent check on current theoretical models. We also note that because the exclusive neutrino reactions tend to be dominated by the form factors at low energy, differential cross sections may be necessary to study the large q^2 behavior of the weak nuclear interaction. Unhappily, these will not be easy to obtain.

REFERENCES

[1] R.C. Allen, Phys.Rev.Letters**64**, 1871(1990).

[2] B.Zeitnitz,Prog. in Part and Nucl. Phys. **13**,445(1995).

[3] T.W. Donnelly, Physics Letters **B43**,93(1973).

[4] S.L. Mintz and M. Pourkaviani,Phys. Rev. **C40**,2458(1989).

[5] M. Fukugita,Y. Kohyama, and K. Kubodera,Physics Lett **B212**, 139(1988).

[6] E. Kolbe,K. Langanke,S. Krewald and F.-K. Thielemann,Nucl. Phys. **A 540**,599 (1992).

[7] S.L. Mintz and M. Pourkaviani,Inclusive and Exclusive Reactions in ^{12}C and Present Experiments,*Proceedings of the 14th International Conference on Particles and Nuclei*,Editors Carl Carlson and John J. Domingo,World Scientific, Singapore,p.587(1997).

[8] M. Pourkaviani and S.L. Mintz,J. Phys. G: Nucl. Part. Phys. **16**,569(1990).

[9] J. Bernabeu and P. Pascual, Nucl. Phys. **A324**,365(1979).

[10] M. Albert et al., Measurement for the Reaction $^{12}C(\nu_\mu, \mu^-)X$ Near Threshold,LAMPF preprint (1994).

[11] S.L. Mintz, and M. Pourkaviani,Nucl. Phys. **A594**,346(1995).

[12] C.W. Kim and S. L. Mintz,Phys. Rev. **C31**,274(1985).

[13] M. Pourkaviani,Lepton Induced Weak Interactions in A = 12 Nuclei and the Weak Nuclear Form Factor,Dissertation,University of Miami ,(1989).

[14] S.L. Mintz,Phys. Rev. **C25**,1671(1982).

[15] C.W. Kim and H. Primakoff,*Mesons in Nuclei*,edited by M. Rho and D.H. Wilkinson,p.68,North Holland, Amsterdam(1979).

[16] Y. Kuno et al.,Z. Phys. **A323**,69(1986).

[17] L. P. Roesch,V.L. Telegdi,P. Truttman, and A. Zehnder, Phys. Rev. Lett. **46**,1507(1981).

[18] W.-Y. Hwang and H. Primakoff,Phys. Rev. **C16**,397(1977).

[19] Y.G.Budyashov et al.,Zh. Eksp. Teor. Fiz.**58**,1211(1970).

[20] G.H. Miller,M. Eckhause,F.R. Kane,P. Martin, and R.E. Welsh, Phys. Lett.**41B**, 50(1972).

[21] M. Giffon et al.,Phys. Rev.**C24**,241(1981).

[22] L.P. Roesch et al.,Phys. Lett.**107B**,31(1981).

[23] C. Tzara, Nucl. Phys. **B18**,246(1970).

[24] S. Nozawa, Y. Kohyama, and K. Kubodera,Prog. Theo. Phys. **70** ,892(1983).

[25] W.C. Louis et al.,Neutrino Oscillation Studies at LAMPF, Preprint,LA-UR 94-2761,1994.

[26] "Neutrino Astrophysics with SNO",A.B.McDonald,Preprint 1994.

[27] S.L. Mintz,M.A. Barnett,G.M. Gerstner, and M. Pourkaviani, Nucl. Phys. **A598** ,367(1995).

[28] J.G. Congleton and H.W. Fearing, Nucl. Phys. **A552**, 534(1993).

[29] J. Frazier and C.W. Kim,Phys. Rev. **177**,2568(1968).

[30] C.W.Kim and H. Primakoff,Phys. Rev. **140**,B566(1965).

[31] C.W. Kim and S.L. Mintz,Phys. Letters **31B**,503(1970).

[32] M. Pourkaviani and S.L. Mintz,J. Phys G: Nucl. Part. Phys.**17**,1139 (1991).

[33] S.L. Mintz and M. Pourkaviani,Nuc. Phys. **A609**,441(1996).

[34] S.L. Mintz and M. Pourkaviani,Phys. Rev. **C37**,2249(1988).

[35] S.L. Mintz and M. Pourkaviani,Phys. Rev. **C47**,873(1993).

SECTION III
PROGRESS ON NEW AND OLD IDEAS

EXOTIC HADRONS

D. B. Lichtenberg

Physics Department
Indiana University
Bloomington, IN 47405, USA

INTRODUCTION

At last year's at Orbis Scientiae I discussed how a broken hadron supersymmetry could be used to obtain predictions about the masses of baryons from knowledge of the masses of mesons.[1] This year, I show how to use hadron supersymmetry to obtain predictions of the masses of some exotic hadrons using as input the masses of ordinary mesons and baryons. In obtaining these results, I do not use any free parameters. However, I do use as input reasonable values of the constituent quark masses taken from a paper on baryons,[2] and I assume that the spin-splitting in ground-state hadrons arises from the chromomagnetic interaction of QCD.[3]

Let me begin with a little history of hadron supersymmetry, first introduced by Miyazawa in 1966,[4,5] well before the proposed supersymmetry of elementary particles. Catto and Gürsey[6,7] showed that Miyazawa's hadron supersymmetry is an approximate consequence of QCD. The supersymmetry arises because a diquark (a boson) and an antiquark (a fermion) are both color (anti)triplets. A diquark can also be in a color-sextet state, but such a state has no connection with a supersymmetry relating diquark to antiquark. From here on, I consider only color-antitriplet diquarks. Some work on hadron supersymmetry is summarized in a review of diquarks.[8]

Because a diquark (a two-quark system considered collectively) and an antiquark are both antitriplets of color SU(3), they both interact similarly with another colored particle. This similarity leads to an approximate invariance under the operation of replacing an antiquark by a diquark. Baryon number is changed under this operation: if an antiquark in a meson is replaced by a diquark, the result is a baryon. Similarly, if a quark and antiquark in a meson are replaced by a diquark and antidiquark, the result is an exotic meson; and if three antiquarks in an antibaryon are replaced by diquarks, the result is a dibaryon. Therefore, to the extent that hadron supersymmetry holds, baryons and exotic mesons can related to mesons, and dibaryons can be related to baryons.

Hadron supersymmetry is broken because a diquark and an antiquark do not have the same mass, spin, or size, and the QCD interaction does depend to some extent on these quantities. I correct for mass and spin differences but not for size differences, remembering that, unlike a current quark, a constituent quark is not a point particle.

Before saying more about the method, I summarize the results. Almost all the exotic mesons of this model have masses well above the threshold for strong decay into ordinary mesons, and so are unlikely to be observed. A possible exception is an exotic containing two b quarks, which may be bound against strong decay, and therefore decay weakly. However, the cross section for the production of a meson with two b quarks is very small. The dibaryons also appear to be unstable against strong decay, including the H dibaryon, composed of $uuddss$ quarks, proposed some time ago.[9]

EXOTIC MESONS

Among the kinds of exotic mesons that may exist are so-called tetraquarks, each composed of a diquark and antidiquark. It should be possible, using broken hadron supersymmetry, to relate the masses of tetraquarks to the masses of mesons.

I neglect spin at first, but later include its effects perturbatively. However, I include the effect of the mass difference between diquark and antiquark from the outset. In the approximation that the force between a quark and an antiquark is the same as the force between an antidiquark and a diquark, I can equate the mass of an exotic meson composed of a diquark and antidiquark to the mass of a meson composed of a fictitious quark and antiquark having the same masses as the diquark and antidiquark.

In order to carry out this procedure, I must obtain the interaction energy of two fictitious quarks having the same masses as the masses of the diquarks, and I must obtain estimates of the diquark masses. This information can come from considering mesons and baryons. First I consider hypothetical spinless mesons and baryons. In some cases, I can estimate the masses of spinless hadrons by averaging over the spins of hadrons with a given quark content. This averaging procedure is the same as the one given in a previous paper.[10] It depends on the assumption that the spin-dependent forces in ground-state hadrons are given by the chromomagnetic interaction of QCD.[3]

In some cases, the spin averaging can be carried out in terms of measured hadron masses.[11] For other hadrons, either not all the relevant spin states have been measured or else the method of ref. 10 is insufficient to eliminate the effects of spin. In these latter cases, I use a semi-empirical mass formula[12] to eliminate the chromomagnetic interaction.

In order to estimate the interaction energy of a bound quark and antiquark with any masses, I first calculate the interaction energy for the known mesons, using as input the following constituent quark masses[2] (in MeV):

$$m_q = 300, \quad m_s = 475, \quad m_c = 1640, \quad m_b = 4985. \tag{1},$$

where the symbol q stands for either u or d. The interaction energies E_{12} are obtained from the simple formula:

$$E_{12} = M_{12} - m_1 - m_2. \tag{2}$$

Because E_{12} is a fairly smooth function of the reduced mass μ of the two particles,[13] if the quark masses are different from the masses of the known quarks I can compute their reduced mass and obtain the interaction energy by interpolation or extrapolation. Because of supersymmetry, I can use this procedure for diquarks, treating them as fictitious (anti)quarks with diquark masses.

I obtain the diquark masses by treating a baryon as a diquark (fictitious quark) bound together with a quark. The interaction energy E_{12} is still given by eq. (2), but now one of the particles, say, particle 2, is a diquark. (I call particle 1 the spectator quark.) If the quarks have different flavors, then I assume that the two heaviest quarks form the diquark.

It is convenient to guess that the diquark mass is the sum of its constituent quark masses. I then compute the reduced mass of the diquark and the spectator quark, obtain the interaction energy from the meson data by interpolation, and so obtain a prediction for the baryon mass. This mass, in general, will not be the same as the observed spin-averaged baryon mass. However, I can then use a revised diquark mass and repeat the process by iteration until the output baryon mass is the same as the actual spin-averaged mass of the baryon. The difference in mass between a diquark and the sum of the masses of the quarks it contains is the triplet interaction energy E_{12}^t. If I use this iteration procedure to calculate the mass of, say, the ss diquark, the result depends on whether the spectator quark in the baryon is q (u or d) or s. The difference is usually less than 20 MeV, and I take the mean of the two values.

Next I have to correct the diquark masses for spin-dependent forces. If the contribution of spin-dependent forces is independent of whether the diquark is in a baryon or in an exotic meson, I can use the values in baryons. These are obtained by assuming that the force arises from the chromomagnetic interaction. The procedure has already been worked out[13] to obtain the spin-dependent contributions either from observed baryon masses or with the help of the semi-empirical mass formula.[12] The resulting diquark masses are given in Table 1.

In general, in an exotic meson there are spin-dependent forces arising between the quarks in the diquark and the quarks in the antidiquark. However, if either the diquark or antidiquark has spin zero, the net contribution of these spin-dependent inter-diquark forces vanishes. For simplicity I consider only exotics with at least one diquark of spin zero. These latter exotics have the lowest masses in any case.

Using the diquark masses in Table 1, I can compute the reduced mass of a system of a diquark and antidiquark. Again interpolating from mesons, I obtain the interaction energy for the diquark-antidiquark state. This procedure gives us the exotic meson mass.

The number of exotic meson masses that can be calculated this way is quite large. A small sample of these exotic masses is shown in Table 2. I also give in Table 2 the lightest mesons into which the exotic can decay strongly, provided the energy permits. The sum of the masses of the decay products are shown in the last column of Table 2.

The calculated exotic masses given in Table 2 have errors of up to 30 MeV associated with the fact that I have chosen the fixed diquark masses of Table 1, whereas actually, in our model, the diquark masses depend mildly on their environment.

It can be seen from Table 2 that only the exotic $qq\bar{b}\bar{b}$ lies below the threshold of the lightest mesons into which it can decay strongly. The exotic $qs\bar{b}\bar{b}$ lies only a little above threshold. It may be observable as a narrow resonance or even a bound state, as our model is probably not accurate enough to distinguish these possibilities in this case. However, exotic mesons with two b quarks are unlikely to be seen for quite some time.

Table 1. Masses M_0 and M_1 of spin-zero and spin-one diquarks obtained from input baryon masses and chromomagnetic interaction energies in baryons. The symbol q stands for u or d, and masses are rounded to the nearest 5 MeV.

Quark content	M_0 (MeV)	M_1 (MeV)
qq	595	800
qs	835	975
ss	–	1150
qc	2100	2150
sc	2250	2295
cc	–	3415
qb	5465	5485
sb	5630	5650
cb	6735	6750
bb	–	10075

DIBARYONS

The deuteron is the only known stable dibaryon, but in principle there might exist others that are stable against strong decay into hadrons. The first such suggestion was made by Jaffe,[9] who proposed that the H dibaryon, composed of the quarks $uuddss$, might be bound against strong decay.

I assume that the H, and its heavier counterparts the H_c ($uuddsc$) and H_b ($uuddsb$) are each composed of three spin-zero diquarks. Although I can obtain only lower limits for dibaryon masses, I can predict that the H and H_c are unstable, but that the H_b might possibly be stable against strong decay. In order for the quarks in each spin-0 diquark to satisfy the Pauli principle, each diquark must contain two quarks of different flavors.

In the limit that diquarks are pointlike, or at least much smaller than the typical separation between different diquarks, I can ignore the Pauli principle for quarks in different diquarks. This is analogous to the case in nuclear physics, in which the quark degrees of freedom in nucleons are generally ignored. In nuclei, when two nucleons are close together, an effective repulsive potential exists between them that arises in part because of the Pauli principle for quarks. Similarly, when two diquarks in a dibaryon are close together, the Pauli principle between quarks of the same flavor in different diquarks should cause an effective short-range repulsion between them. In order to use supersymmetry to relate the diquark-diquark interaction to the quark-quark interaction, I have to neglect this repulsion, thereby underestimating the mass of a dibaryon containing three diquarks.

Next I combine any two diquarks into a color-triplet quadriquark. I can obtain the mass of the quadriquark by adding the appropriate interaction energy E_{12}^t to the sum of the masses of the diquarks, taken from Table 1. The mass of a dibaryon is obtained by adding the appropriate interactions energy E_{12} to the sum of the masses of the quadriquark and diquark Of all possible ways to choose the quadriquark from two out of three diquarks, I pick the way that yields the lowest dibaryon mass, as I am looking for a lower limit.

Table 2. Predicted masses M_E of some exotic mesons obtained from the diquark masses in Table 1. These are all ground-state mesons with the given quark content, and the parities are all positive. The symbol q stands for u or d, and masses are rounded to the nearest 10 MeV. The two numbers in column 2 are the spin of the diquark and antidiquark respectively; the spin of the exotic is just the sum of these spins. Thus, all these exotics are either $J^P = 0^+$ or $J^P = 1^+$. The next-to-last column gives the lowest-mass mesons into which the exotic can decay strongly if energetically permitted, and the last column gives the threshold energy E_t of the decay.

Quark content	Diquark spins	M_E (MeV)	Decay products	E_t (MeV)
$qq\bar{q}\bar{q}$	0, 0	1180	$\pi\pi$	280
$qq\bar{q}\bar{q}$	0, 1	1370	$\pi\rho$	910
$qq\bar{q}\bar{s}$	0, 0	1400	πK	630
$qq\bar{q}\bar{s}$	1, 0	1580	πK^*	1030
$qq\bar{s}\bar{s}$	0, 1	1700	KK^*	1390
$qs\bar{q}\bar{s}$	0, 0	1610	$K\bar{K}$	990
$qs\bar{q}\bar{s}$	0, 1	1740	$K^*\bar{K}, K\bar{K}^*$	1390
$qq\bar{q}\bar{c}$	0, 0	2620	$\pi\bar{D}$	2010
$qq\bar{q}\bar{c}$	0, 1	2660	$\pi\bar{D}^*$	2150
$qq\bar{q}\bar{c}$	1, 0	2770	$\pi\bar{D}^*$	2150
$qq\bar{q}\bar{b}$	0, 0	5950	πB	5420
$qq\bar{q}\bar{b}$	0, 1	5970	πB^*	5460
$qq\bar{s}\bar{c}$	0, 0	2760	$K\bar{D}$	2360
$qs\bar{q}\bar{c}$	0, 0	2800	$\pi\bar{D}_s$	2110
$qq\bar{c}\bar{c}$	0, 1	3910	$\bar{D}\bar{D}^*$	3880
$qc\bar{q}\bar{c}$	0, 0	3920	$\pi\eta_c$	3120
$qq\bar{c}\bar{b}$	0, 0	7220	$\bar{D}B$	7150
$qc\bar{q}\bar{b}$	0, 0	7180	πB_c	6390
$qs\bar{c}\bar{b}$	0, 0	7380	$\bar{D}B_s$	7240
$qq\bar{b}\bar{b}$	0, 1	10550	BB^*	10600
$qb\bar{q}\bar{b}$	0, 0	10380	$\pi\eta_b$	9540
$qs\bar{b}\bar{b}$	0, 1	10710	BB_s^*	10700

The mass of the H dibaryon satisfies

$$M(H) \geq 2320 \text{ MeV}, \tag{3}$$

rounded to the nearest 10 MeV. This mass is about 90 MeV above the mass of two Λ baryons, and so the model says that the H can decay strongly into two Λ's. Because there is nothing to inhibit the decay, the decay width may be too large to make the state readily observable. Similarly, the mass of the H_c dibaryon satisfies

$$M(H_c) \geq 3460 \text{ MeV}, \tag{4}$$

a value 60 MeV above threshold for decay into $\Lambda + \Lambda_c$. Again, the decay width might be too large to allow the state to be observable. Lastly, the mass of the H_b satisfies

$$M(H_b) \geq 6730 \text{ MeV}, \tag{5}$$

This last value is 10 ± 20 MeV *below* the threshold for decay into $\Lambda + \Lambda_b$, but of course this is only a lower limit. Nevertheless, this dibaryon is well worth searching for. If it is bound, it should decay weakly into the two-body final state $\Lambda + \Lambda_c$ as well as into other hadrons including a charmed hadron.

DISCUSSION

The model has one important advantage and one important drawback. The advantage is I can predict the masses of certain exotic hadrons from supersymmetry from the properties of ordinary mesons and baryons and constituent quark masses without needing any free parameters. The drawback is that because I do not use a Hamiltonian, I do not know how to obtain wave functions and decay rates. However, I see no reason why any energetically allowed decays should be inhibited, and therefore estimate these decay widths to be large.

I stress that the model of exotic hadrons restricts them to be composed of color-antitriplet diquarks. This restriction is necessary to use supersymmetry to predict the masses of exotic mesons and lower limits to the masses of dibaryons without using any free parameters. However, the actual structure of the lowest-mass exotic hadrons may be quite different from what I have assumed.

In a recent work, Pepin et al.,[14] using a potential model, find that exotic mesons containing two heavy (c or b) quarks are bound by 200 to 400 MeV. An important feature of their work is the assumption that the spin-spin interaction between quarks does not arise primarily from the chromomagnetic interaction of QCD but from pweudoscalar-meson exchange. Of course, if I were to assume, like these authors, that the spin-spin interaction depends on flavor rather than color, my results would be different.

Turning to dibaryons, Buchmann et al.[15] have recently used a potential model to calculate the mass of a dibaryon containing only u and d quarks. They find that the Pauli principle can raise the mass of a dibaryon made of three diquarks by more than 100 MeV. If the effect of the Pauli principle is as large in H_b, then it is not bound in my model.

In conclusion, broken hadron supersymmetry leads to predictions the masses of exotic hadrons from the masses of ordinary mesons and baryons. Most exotic mesons (except for the $qq\bar{b}\bar{b}$ and possibly the $qs\bar{b}\bar{b}$) are unbound against strong decay. Likewise, the H and H_c dibaryons should be able to decay strongly, but whether the H_b is stable against strong decay depends to a large extent on effects of the Pauli principle that have been neglected.

Acknowledgments

Most of this work has been done in collaboration with Enrico Predazzi and Renato Roncaglia. This work was supported in part by the U.S. Department of Energy.

REFERENCES

1. D. B. Lichtenberg, Hadron supersymmetry and relations between meson and baryon masses, in: *Neutrino Mass, Dark Matter, Gravitational Waves, Monopole Condensation, and Light Cone Quantization*, B.N. Kursunoglu, S. L. Mintz, and A. Perlmutter, eds. Plenum, New York (1996).

2. R. Roncaglia, D. B. Lichtenberg, and E. Predazzi, Predicting the masses of baryons containing one or two heavy quarks, *Phys. Rev. D* 52:1722 (1995).

3. A. Re Rújula, H. Georgi, and S. L. Glashow, Hadron masses in a gauge theory, *Phys. Rev. D* 12:147 (1975).

4. H. Miyazawa, Baryon number changing currents, *Prog. Theor. Phys.* 36:1266 (1966).

5. H. Miyazawa, Spinor currents and symmetries of baryons and mesons, *Phys. Rev.* 170:1586 (1968).

6. S. Catto and F. Gürsey, Algebraic treatment of effective supersymmetry, *Nuovo Cimento* 86:201 (1985).

7. S. Catto and F. Gürsey, New realizations of hadronic supersymmetry, *Nuovo Cimento* 99:685 (1988).

8. M. Anselmino, E. Predazzi, S. Ekelin, S. Fredriksson, and D.B. Lichtenberg, Diquarks, *Rev. Mod. Phys.* 65:1199 (1993).

9. R. L. Jaffe, Perhaps a stable dihyperon, *Phys. Rev. Lett.* 38:195 (1977).

10. M. Anselmino, D.B. Lichtenberg, and E. Predazzi, Quark color-hyperfine interactions in baryons, *Z. Phys. C* 48:605 (1990).

11. R. M. Barnett et al., Review of particle properties, *Phys. Rev. D* 54:1 (1996).

12. Y. Wang Y and D. B. Lichtenberg, Semiempirical formulas for the color-hyperfine mass splittings in hadrons, *Phys. Rev. D* 42:2404 (1990).

13. R. Roncaglia, A. Dzierba, D. B. Lichtenberg, and E. Predazzi, Predicting the masses of heavy hadrons without an explicit Hamiltonian, *Phys. Rev. D* 51:1248 (1995)

14. S. Pepin, Fl. Stancu, M. Genovese, and J.-M. Richard, Multiquark systems in a constituent quark model with chiral dynamics, nucl-th/9608058 and ph-9609348 (unpublished).

15. A. J. Buchmann, G. Wagner, K. Tsushima, L. Ya. Glozman, and A. Faessler, The d'-dibaryon in the nonrelativistic quark model, *Prog. Part. Nucl. Phys.* 36:383 (1996).

ORTHOGONAL MIXING AND CP VIOLATION

Paul H. Frampton

Department of Physics and Astronomy
University of North Carolina
Chapel Hill, NC 27599-3255

I. INTRODUCTION

In this talk I discuss (I) the history of CP, (II) the aspon model, (III) production of A and Q in pp collisions at LHC (IV) B decay (V) Cabibbo mixing and finally (VI) summarize.

The parity operation $\underline{x} \rightarrow -\underline{x}$ is an invariance of Newton's laws with the strong form of the 3rd Law, that action and reaction act along the line of centers. In quantum mechanics, the P operator was introduced by Wigner [1] in 1927. Interactions between fields fix relative parities - P is a good symmetry for strong and electromagnetic interactions.

P violation was first seriously entertained in Lee and Yang [2] and first shown experimentally in C.S. Wu et al. [3] [Note: Madame Wu died the month after this conference, on February 16, 1997.], also in Lederman et al., and Telegdi et al. [4]

Time reversal, $t \rightarrow -t$ is an invariance of Newton's laws. For quantum mechanics, T invariance was first studied by Wigner in 1932 [5] who introduced the familiar anti-unitary operator

$$T\left(a|\psi> + b|\phi>\right)$$
$$= a^* \, T|\psi> + b^* \, T|\phi>$$

[T invariance violation was studied in classical statistical mechanics earlier by Boltzmann, but T violation in microscopic laws was not seriously in question until 1964].

Charge conjugation (C) which takes matter → antimatter had to wait until at least 1928 and the Dirac equation which predicted e^+, discovered in 1932, and $\bar{p}$ discovered in 1955. C invariance of quantum electrodynamics was first discussed by Kramer [6] in 1937. In Dirac theory, C is some 4 x 4 matrix (M)

$$\psi \xrightarrow{C} M\psi^*$$

The invariance under CPT was proven for quantum field theory in 1954 by Lüders [7] under weak assumptions of lorentz invariance and the spin-statistics connection.

After Lee and Yang, but before P violation was discovered, Landau [8] suggested that CP is an exact symmetry.

In [9] CP violation was discovered in the decay of $K^o(S = +1)$ and $\overline{K^o}(S = -1)$ produced by strong interactions. The weak decays which conserve CP are $K_S \to \pi^+\pi^-$ (CP = +) and $K_L \to \pi^o \pi^o \pi^o$ (CP = −). But $K_L \to \pi \pi$ was observed 0.2% of the time. Defining

$$K_{\substack{S \\ L}} = \frac{(1 + \varepsilon) \mid K^o> \pm (1 - \varepsilon) \mid \overline{K^o}>}{\sqrt{2(1 + \mid \varepsilon \mid^2)}}$$

one has $\varepsilon \approx 0.002$.

CP violation has not been seen outside the $K^o - \overline{K}^o$ system. The parameter $\mathrm{Re}(\varepsilon'/\varepsilon)$ which measures the amount of $\Delta I = 3/2$ CP violation is still uncertain experimentally being measured at $\sim 6 \times 10^{-4}$ at Fermilab and $\sim 23 \times 10^{-4}$ at CERN. So it is really only one parameter, ε, which is well-known to 1% accuracy in CP violation.

In a remarkable paper, Sakharov in [10] proposed that the net baryon number of the universe arose due to a combination of three ingredients:

(i)   B violating interactions
(ii)   Thermodynamic out-of-equilibrium
(iii)   CP violation

When GUTs became popular Yoshimura [11] and others illustrated this idea. More recently, electroweak baryogenesis is discussed using (i), (ii) and (iii).

In 1973 Kobayashi and Maskawa [12] proposed their KM mechanism. In the mixing of 2N quark flavors there is an N x N unitary matrix with $N^2 - (2N - 1) = (N - 1)^2$ parameters; the $(2N - 1)$ are unphysical relative phases. There are $\frac{1}{2}N(N - 1)$ rotation angles and the remainder, if any, are CP violating phases. For N = 2 there is one angle, the Cabibbo angle. For N = 3 there are three angles and a phase. Only u,d,s were established in 1973. With c,t,b now established this source of CP violation must be present in the Standard

Model (SM). It is not known whether it is the only source, a principal source, or a small contributor to $K^o - \overline{K}^o$ CP violation.

The KM mechanism is by far the most popular, being the most conservative, for CP violation. According to SPIRES, the KM paper has 2500 citations, the third most of all particle theory papers.

In 1975/76 with the advent of instantons 't Hooft [13] and others emphasized the strong CP problem. A serious difficulty with QCD is that a term $\sim \overline{\theta} \, G_{\mu\nu}\widetilde{G}^{\mu\nu}$ in L_{QCD} requires fine-tuning to $\overline{\theta} < 10^{-10}$ in order to be consistent with the upper limit on the neutron electric dipole $d_n < 10^{-25}$ e cm. Possible solutions include the $U(1)_{PQ}$ axion which is ruled out, although the invisible-axion modification is not yet completely excluded.

In the decade 1980-89 the areas of weak CP and strong CP proceeded along largely separate tracks. In the rest of the talk I will discuss papers: two [14, 15] in 1991, one [16] in 1992, one [17] in 1994 and finally one [18] in 1997 which gives a new motivation for studying this model. There will be some brief discussion from each of the five papers.

II. THE ASPON MODEL

As already mentioned, QCD has a term $\overline{\theta}_{QCD} \, G \, \widetilde{G}$ violating CP (and P). Physical effects depend on $\overline{\theta} = \left(\theta_{QCD} + \theta_{QFD}\right)$ where θ_{QFD} is the argument of the determinant of the quark mass matrix.

There are two approaches to strong CP: a color anomalous $U(1)_{PQ}$ which allows $\overline{\theta}$ to go to zero, or imposing CP symmetry on the theory which gives $\theta_{QCD} = 0$ and then arranging $\theta_{QFD} = 0$ at lowest order. The first approach gives rise to many additional questions requiring further research. But axion searches remain frustrated so we may look at the second approach for testable physics.

If we write the (T_3, Y) quantum numbers for one generation of quarks, they are

$$\left(-\tfrac{1}{2}, \tfrac{1}{6}\right) d_L \quad \left(0, \tfrac{1}{3}\right) \overline{d}_L$$
$$\left(\tfrac{1}{2}, \tfrac{1}{6}\right) u_L \quad \left(0, -\tfrac{2}{3}\right) \overline{u}_L$$

Under a new gauge group $U(1)_{new}$ we assign $Q_{new} = 0$ for these quarks, and also the leptons. Similarly the second and third families, as well as the $\left(\tfrac{1}{2}, -\tfrac{1}{2}\right)$ Higgs doublet, have $Q_{new} = 0$. Additional states are a real representation of quarks (U,D) with $\left(\pm\tfrac{1}{2}, \tfrac{1}{6}\right) + \left(\pm\tfrac{1}{2}, -\tfrac{1}{6}\right)$ and $Q_{new} = \pm 1$ and a pair of complex scalar singlets $\chi_{1,2}$ (0,0) having $Q_{new} = +1$.

The Yukawa couplings are the SM ones $\overline{u}_L^i u_L^j \phi$, etc. together with new ones $h_i^\alpha U_L \, u_L^i \, \chi_\alpha$, etc. The VEVs of χ_α are $\langle\chi_1\rangle = k_1$, $\langle\chi_2\rangle = k_2 \, e^{i\theta}$ spontaneously breaking $U(1)_{new}$ and CP. The gauge boson becomes massive by the Higgs mechanism and was called the "aspon" in 1990. The 4 x 4 quark mass matrix has complex entries but _it_ has a real determinant. Together with the assumption of CP asymmetry this ensures that $\theta = 0$ at the tree level. The mass matrix has the texture [19]

$$\begin{pmatrix} m_{ij} & F_i \\ O & M \end{pmatrix}$$

where M is the Dirac mass of the new quarks.

Putting, for example, the down-quark matrix in the canonical form

$$M_{down} = \begin{pmatrix} m_d & O & O & F_1 \\ O & m_s & O & F_2 \\ O & O & m_b & F_3 \\ O & O & O & M \end{pmatrix}$$

where $F_i = \sum\limits_{\alpha} h_i^{\alpha} (\chi_\alpha)$ this can be diagonalized by the bi-unitary transformation

$$K_L^t \, M_{down} \, K_R = \text{diag}\left(M_d', M_s', M_b', M'\right)$$

Similarly

$$J_L^t \, M_{up} \, J_L = \text{diag}\left(M_u', M_c', M_t', M'\right)$$

The matrices $K_{L,R}$ and $J_{L,R}$ can be found explicitly to order x^2 where $x_i = F_i/M$ are small parameters $|x_i| \ll 1$. We sometimes suppress the family label and generically write x for x_i.

To quadratic order K_L is given by

$$K_L = \begin{pmatrix} 1 - \frac{1}{2}|x_1|^2 & \dfrac{x_1 \, x_2^* \, m_s^2}{m_d^2 - m_s^2} & \dfrac{x_1 x_3^* m_b^2}{m_d^2 - m_b^2} & x_1 \\[2ex] \dfrac{x_1^* x_2 \, m_d^2}{m_s^2 - m_d^2} & 1 - \frac{1}{2}(x_2)^2 & \dfrac{x_2 \, x_3^* \, m_b^2}{m_s^2 - m_b^2} & x_2 \\[2ex] \dfrac{x_1^* x_3 \, m_d^2}{m_b^2 - m_d^2} & \dfrac{x_2^* \, x_3^* \, m_s^2}{m_b^2 - m_s^2} & 1 - \frac{1}{2}(x_3)^2 & x_3 \\[2ex] -x_1^* & -x_2^* & -x_3^* & 1 - \frac{1}{2}\Sigma x_i^2 \end{pmatrix}$$

Flavor-Changing Neutral Currents (FCNC) are induced in the right-handed sector since we have violated the rule that all right-handed quarks of the same electric charge must have the same T_3. The induced FCNC are given by:

$$-\frac{1}{2}\frac{g_2}{\cos\theta_W}\,\overline{D}_R\,\gamma_\mu\,D_R\,Z^\mu$$
$$= \beta_{ij}\,\overline{d}_R^{\prime i}\,\gamma_\mu\,d_R^{\prime j}\,Z^\mu \quad (i \neq j)$$

with

$$\beta_{ij} = \frac{1}{2}\frac{g_2}{\cos\theta_W}\frac{m_{di}\,m_{dj}}{M^2}\,x_i x_j^*$$

From this one finds $\beta_{12} < 10^{-10}$ while $\beta_{12} < 10^{-6}$ is all that is required for phenomenology.

At one-loop level $\overline{\theta}$ acquires a non-zero contribution from the diagram

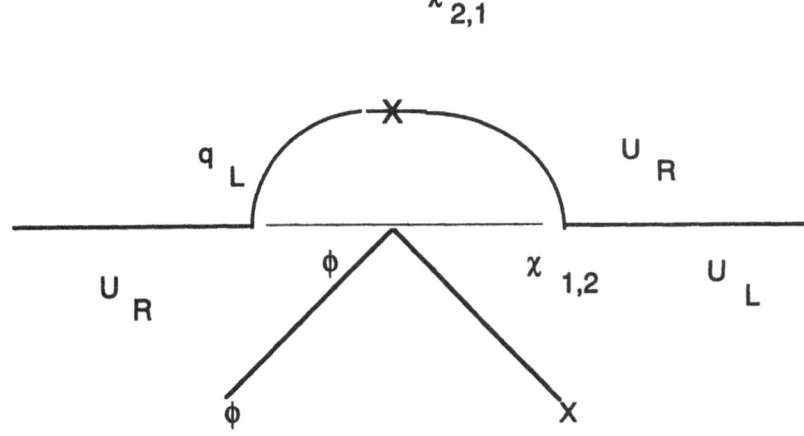

Using

$$\overline{\theta}(up) = \text{Arg}\left[\det\left(M_{up} + \delta M_{up}\right)\right]$$
$$= \text{Im Tr} \ln\left[M_{up}\left(1 + M_{up}^{-1}\delta M_{up}\right)\right]$$
$$\approx \text{Im Tr}\left(M_{up}^{-1}\,\delta M_{up}\right)$$

The general result for $\overline{\theta}$ is

$$\overline{\theta} \approx \frac{1}{16\pi^2}\lambda x^2$$

where λ is the coefficient in $\lambda |\phi|^2 |\chi|^2$. This then puts constraints on the allowed values for λ and x^2 which will be used below.

Fitting to the CP violating parameter $\varepsilon = 2.26 \times 10^{-3}$ and $\left| \text{Re}(\varepsilon'|\varepsilon) \right| < 4 \times 10^{-3}$ gives the upper limit on the scale of breaking $U(1)_{new}$, namely $\kappa_{1,2} \leq 2$ TeV. Correlated to this, one expects that the new particles A and Q should have masses $M(Q) \leq 600$ GeV and $M(A) \leq 600$ GeV, and hence be accessible to the LHC.

It is possible to obtain an intuition about such upper limits. A principal contributing diagram is

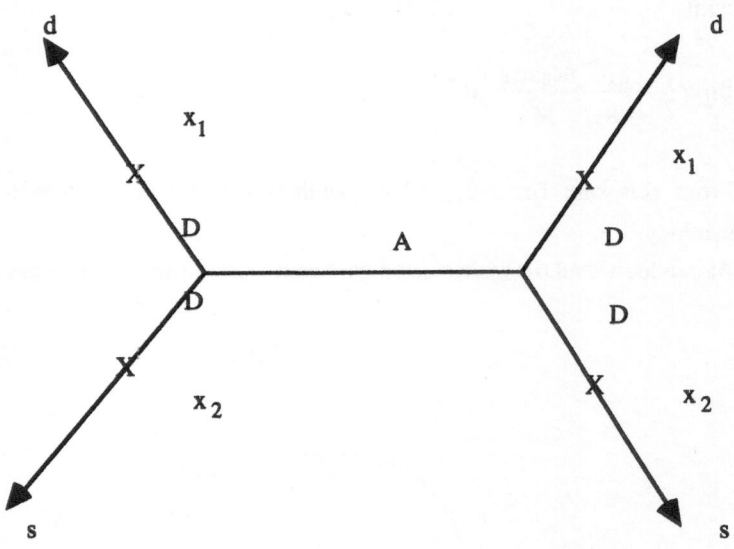

The x_i have upper limits from the calculation of $\bar{\theta}$. If $M(Q) = M(D)$ and/or $M(A)$ are too large, then the diagram becomes too small to account for ε.

III. PRODUCTION OF A AND Q IN pp AT LHC

The production of $Q\bar{Q}$ is dominated by gluon fusion diagrams

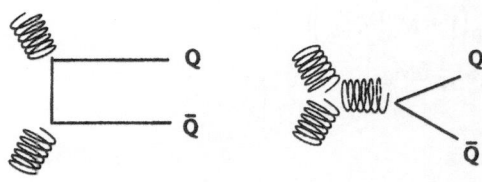

just like e.g. $t\bar{t}$ production. The cross section for $\sqrt{s} = 14$ TeV falls gently with $M(Q)$:

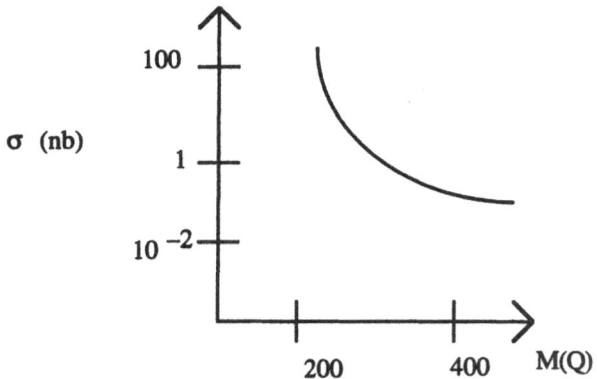

Production of A is by bremsstrahlung from the heavy quark

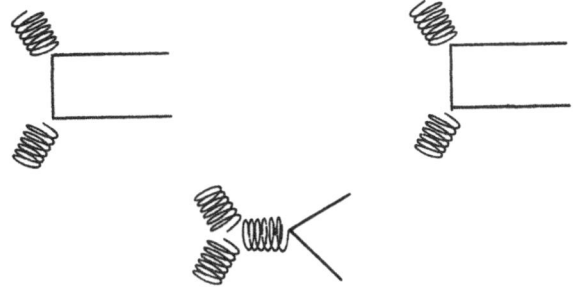

The resultant cross-section for M(Q) = 100, 300 GeV is given by

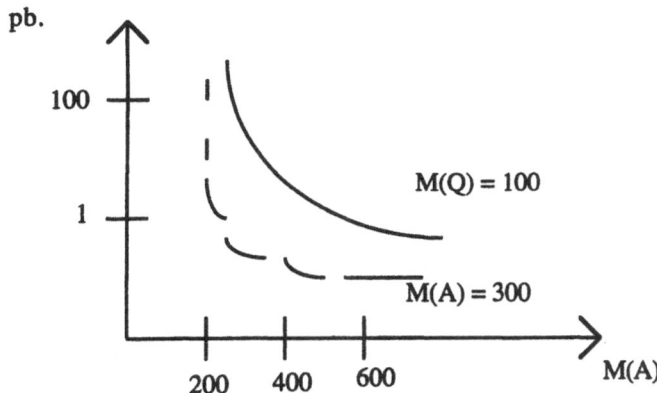

The decay width of A depends on the availability of decay channels into the exotic quarks. If $M(A) > 2M(Q)$ one finds $\Gamma_A \sim 1\text{GeV}$ while for $2M(Q) > M(A) > M(Q)$, $\Gamma_A \sim 1\text{MeV}$ and for $M(A) < M(Q)$, $\Gamma_A \sim 1\text{keV}$. The dominant decay modes are $A \rightarrow Q\overline{Q}$, $A \rightarrow Q\overline{q}$, $A \rightarrow q\overline{q}$ respectively for these three ranges. Clearly, A can be an extraordinarily-long lived state for such a heavy particle! Production of $\left(Q\overline{Q}\right)$ - onia has been analyzed also [16].

The key result for production is that the cross-sections are large enough to give more than an adequate number of events at the LHC.

IV. B DECAY

The KM mechanism of CP violation can be nicely checked by the unitarity triangle whose three sides are the terms in

$$V_{ub}^* \, V_{ud} + V_{tb}^* \, V_{td} + V_{cb}^* \, V_{cd} = 0.$$

The angles α, β, γ are between the (1st, 2nd), (2nd, 3rd), (3rd, 1st) terms respectively, as follows:

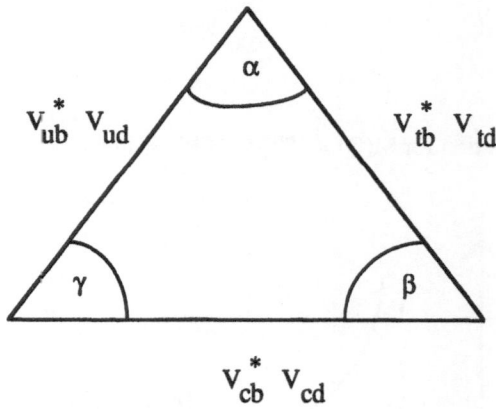

It is useful to rescale the triangle by dividing everything with $V_{cd}^* \, V_{cb}$. This gives a triangle with vertices corresponding to the Wolfenstein parametrization [20] of V_{CKM} $(0,0)$, $(1,0)$ and (ρ, η) in a ρ-η plot.

$$V_{CKM} = \begin{pmatrix} 1 & \lambda & A\lambda^3 (\rho - i\eta) \\ -\lambda & 1 & A\lambda^2 \\ A\lambda^3 (1-\rho-i\eta) & -A\lambda^2 & 1 \end{pmatrix}$$

In the B factories (in Tokyo and at SLAC) the principal aim is to measure the sides and angles of this unitarity triangle, i.e. to determine $(\rho$-$\eta)$ of the upper vertex. If we assume the KM mechanism for the $K^\circ - \overline{K}^\circ$ CP violation then $\eta \neq 0$ and appreciable CP asymmetry is observable in B decay.

In an asymmetric e^+e^- collider, one measures:

$$a_f(t) = \frac{\Gamma(B^0(t) \to f) - \Gamma(\overline{B}^0(t) \to f)}{\Gamma(B^0(t) \to f) + \Gamma(\overline{B}^0(t) \to f)}$$

We define:

$$|B_{1,2}> = p \, | \, B^\circ > \pm q \, | \, \overline{B}^\circ >$$
$$A, \overline{A} = < f \, | \, H \, | \, B^\circ, \overline{B}^\circ >$$

where f = CP eigenstate. $\lambda(f)$ is defined by

$$\lambda(\pi^+\pi) = \left(\frac{q}{p}\right)_{B_d} \left(\frac{\overline{A}}{A}\right)_{B_d \to \pi^+\pi^-}$$

$$\lambda(\psi \, K_s) = \left(\frac{q}{p}\right)_{B_d} \left(\frac{\overline{A}}{A}\right)_{B_d \to \psi K} \left(\frac{q}{p}\right)_K$$

$$\lambda(\rho K_s) = \left(\frac{q}{p}\right)_{B_s} \left(\frac{\overline{A}}{A}\right)_{B_s \to \eta K} \left(\frac{q}{p}\right)_K$$

The measured CP asymmetries are

$$a_f(t) = -\,\text{Im}\,\lambda(f)\,\sin(\Delta Mt)$$

where ΔM is the $B_1 - B_2$ mass difference.

For the KM case for which this formalism was developed:

$$\sin 2\alpha = \text{Im}\,\lambda(\pi^+\pi^-)$$

$$\sin 2\beta = \text{Im}\,\lambda(\psi \, K_s)$$

$$\sin 2\gamma = -\,\text{Im}\,\lambda(\rho \, K_s)$$

The $\psi \, K_s$ mode is called "gold plated" because it is easiest to measure. All of these quantities are $0(1)$ in KM. In the present model, explicit calculation gives

$$\left| \text{Im}\,\lambda(\pi^+\pi^-) \right| \leq 1 \times 10^{-5}$$

$$\left| \text{Im } \lambda(\psi K_s) \right| \leq 2 \times 10^{-3}$$

$$\left| \text{Im } \lambda(\rho K_s) \right| \leq 2 \times 10^{-3}$$

so in this case β, γ are tiny angles, a milliradian or less, and α is within a whisker of $\alpha = \pi$.

In the present model, CP asymmetry is too small to be detectable in the planned B factories.

There are specific predicted relations which can eventualy be checked e.g.

$$\text{Im } \lambda(\psi_s) + \text{Im } \lambda(k_S)$$
$$- \text{Im } \lambda(\pi^+\pi^-) - \text{Im } \lambda(D_s^+ D_s^-) = 0.$$

V. CABIBBO MIXING

A re-analysis and new experimental input recently led to a renaissance of this model and its "preliminary confirmation."

Recall that for $K^0 - \overline{K}^0$ CP violation arises not from the $W^+ W^-$ box diagram but from the alternative box:

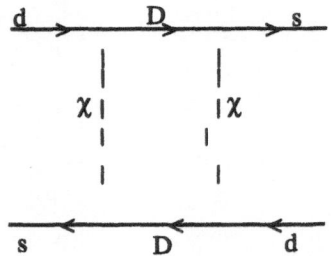

A calculation of this gives $x^2 \geq 3 \times 10^{-5}$. From the one-loop calculation of $\overline{\theta}$ we know that

$$\overline{\theta} \approx \frac{\lambda x^2}{16\pi^2} \leq 10^{-10}$$

so that $\lambda x^2 < 10^{-8}$. Bearing in mind the upper limit on x^2 from $\overline{\theta}$ and applying naturalness considerations to the coupling $\lambda |\phi|^2 |x|^2$, using $m_H > 60$ GeV and $\langle \chi \rangle < 2$TeV we estimate that $10^{-3} > x^2 > 3 \times 10^{-5}$ is the allowed window for x^2 in the aspon model.

This small quantity x^2 measures the departure from real orthogonality of the 3 x 3 quark mixing matrix. In such Cabibbo mixing for three families consistent with B decay data? Recall that the sides of the unitarity triangle have lengths:

$$1, \quad R_b = \left| \frac{V_{ud}^* V_{ub}}{V_{cd}^* V_{cb}} \right|, \quad R_t = \left| \frac{V_{cd}^* V_{tb}}{V_{cd}^* V_{cb}} \right|$$

The data support $R_b = 0.35 \pm 0.09$. Using KM one can derive that $R_t = 0.99 \pm 0.22$ [21]. This corresponds to $0.34 \leq \sin 2\beta \leq 0.75$, a large CP violating angle.

On the other hand, if there is orthogonality up to order $x^2 \leq 10^{-3}$ then we predict $R_b + R_t = 1 \pm 0.001$. We can independently extract $R_t = 1 - p$ to deduce that $R_t = 0.637 \pm 0.09$ which implies $R_b + R_t = 0.99 \pm 0.13$. It is non-trivial that the present model is consistent with data at the 13% level, and that the data allow $\beta, \gamma \approx 0$.

VI. SUMMARY

Only one CP violation parameter has been measured with any precision:

$|\varepsilon| = 0.002263 \pm 0.000023$

$\arg(\varepsilon) = 43.5 \pm 1.0°$

and this fact makes it impossible at present to pin down the correct theory.

The KM mechanism must be present at some level, but it does not address - as does not the standard model - the strong CP problem.

The present model which does solve strong CP has definite predictions:

(i) $R_b + R_t = 1.000 \pm 0.001$. This is consistent with the present data 0.99 ± 0.13. This can be checked further at CLEO, and will certainly be checked at B Factories at SLAC and KEK in 1999 and beyond.

(ii)   The narrow particle (aspon) will be produced by $gg \rightarrow Q\bar{q}A$ at LHC in 2005 and beyond. We predict $m(A) < 600$ GeV.

The issue of CP violation is important because it is why we exist. My pennies are on a spontaneous CP breaking because it solves the strong CP problem, and is consistent with present data.

ACKNOWLEDGEMENT

This work was supported in part by the U.S. Department of Energy under Grant No. DE-FG05-85ER-40219.

REFERENCES

1. E. P. Wigner, *Physik* 43, 624 (1927)
2. T.D. Lee and C.N. Yang, *Phys. Rev.* 104, 254 (1956)
3. C.S. Wu et al., *Phys. Rev.* 105, 1413 (1957)
4. L. Lederman et al., ibid 105, 1415 (1957), V. Telegdi et al., ibid 105, 1681 (1957)
5. E.P. Wigner, *Gott. Nachr.* 546 (1932)
6. H.A. Kramers, *Proc. Acad. Amst.* 40, 814 (1937)
7. G. Luders, *Kg. Dansk. Vidersk. Selsk. Mat.-Fys. Medd.* 28 #5 (1954)
8. L.D. Landau, *Nucl. Phys.* 3, 127 (1957)
9. J.H. Christenson, J.W. Cronin, V.L. Fitch and R. Turlay, *Phys. Rev. Lett.* 13, 138 (1964)
10. A.D. Sakharov, *JETP Letters* 5, 24 (1967)
11. M. Yoshimura, *Phys. Rev. Lett.* 41, 281 (1978)
12. M. Kobayashi and T. Maskawa, *Prog. Theor. Phys.* 49, 652 (1973)
13. G. 't Hooft, *Phys. Rev. Lett.* 37, 8 (1976)
14. P.H. Frampton and T.W. Kephart, *Phys. Rev. Lett.* 66, 1666 (1991)
15. P.H. Frampton and D. Ng, *Phys. Rev.* D43, 3034 (1991)
16. P.H. Frampton, T.W. Kephart, D. Ng and T.J. Weiler, *Phys. Rev. Lett.* 68, 2129 (1992)
17. A.W. Ackley, P.H. Frampton, B. Kayser and C.N. Leung, *Phys. Rev.* D50, 3560 (1994)
18. P.H. Frampton and S.L. Glashow, *Phys. Rev.* D55, 1691 (1997)
19. See also A. Nelson, *Phys. Lett.* 136B, 387 (1984); S.M. Barr, *Phys. Rev.* D30, 1005 (1984)
20. L. Wolfenstein, *Phys. Rev. Lett.* 51, 1945 (1983)
21. M. Neubert, *Int. J. Mod. Phys.* A11, 4173 (1996)

CPT VIOLATION, STRINGS, AND NEUTRAL-MESON SYSTEMS

V. Alan Kostelecký

Physics Department
Indiana University
Bloomington, IN 47405
U.S.A.

INTRODUCTION

Symmetry under the discrete transformation CPT is a general theoretical condition holding for local relativistic field theories of point particles [1]-[7]. This symmetry has been investigated experimentally under different circumstances and to a high degree of precision [8]. The broad theoretical validity of CPT symmetry for particles and the availability of high-precision tests makes CPT violation an interesting candidate experimental signal for fundamental theories such as string theory [9, 10, 11].

In this talk, I consider the possibility that CPT symmetry might be violated in nature by effects arising in a theory beyond the standard model. One example is string theory, which presently provides the most promising framework for a consistent quantum theory of gravity incorporating also the known interactions and particles. Strings are extended objects, so the usual assumptions underlying proofs of CPT symmetry do not hold. Indeed, spontaneous CPT violation can occur in string theory [9, 10], via a mechanism outlined in the next section.

If physics beyond the standard model includes spontaneous CPT violation, it can be described at low energy by additional terms in an effective theory. It is possible to establish the general form of such terms that are compatible with known gauge symmetries [11, 12]. This analysis in turn suggests possible consequences of CPT violation such as baryogenesis [13] and, in particular, quantitative experimental tests of CPT. Among the most promising tests are those involving neutral-meson oscillations, where specific signatures appear in experiments involving either correlated or uncorrelated mesons. These are oulined briefly for the various neutral-meson systems in subsequent sections. Further details about these effects and experiments can be found in the original literature on the K system [9, 10, 11], the two B systems [11, 14, 15], the D system [11, 16].

The possible spontaneous CPT violations discussed here, which are tied to minuscule spontaneous violations of Lorentz invariance [17], lie entirely within the framework of conventional quantum mechanics. It has also been suggested [18]-[20] that violations of conventional quantum mechanics possibly arising in the context of quantum gravity might lead to CPT breaking. The experimental signatures of the two types of CPT violation in the kaon system are entirely distinct [21].

SPONTANEOUS CPT VIOLATION

If a fundamental theory underlying nature involves more than four spacetime dimensions and is dynamically Poincaré invariant, some type of spontaneous breaking of the higher-dimensional Poincaré group presumably occurs to generate a four-dimensional effective theory. String theory is most naturally formulated in higher dimensions and indeed has a mechanism that can trigger spontaneous Lorentz violation [17]. In string field theory, this mechanism involves certain interactions that do not appear in conventional four-dimensional renormalizable gauge theories. The string gauge invariance admits these interactions as a consequence of string nonlocality or, equivalently, as a consequence of the appearance of an infinite number of particle fields. If scalar fields in the string theory acquire vacuum expectation values, these interactions can cause destabilizing effects on the static potentials for Lorentz tensor fields. A stable vacuum can then become one in which Lorentz tensor fields have nonzero expectation values, thereby spontaneously breaking Lorentz invariance. If these tensors include ones with an odd number of spacetime indices, the spontaneous Lorentz breaking also involves CPT breaking.

The string field theory of the open bosonic string provides a useful explicit testing ground for these ideas. A level-truncation scheme can be used to explore the space of extrema for the action in a systematic way. The idea is to construct the action and the equations of motion analytically, using all particle fields up to a given level number. The solutions to the equations of motion that break Lorentz and CPT invariance can be found and compared with similar solutions for truncations at different level numbers. Solutions of interest are those that, as the level number is increased, both persist and are corrected by smaller and smaller amounts. For some situations, symbolic-manipulation techniques have enabled us to treat over 20,000 nonvanishing terms in the action. The Lorentz and CPT properties expected from the theoretical mechanism agree with those of the solutions found via the level-truncation approach.

CPT-VIOLATING EXTENSION TO THE STANDARD MODEL

An interesting issue is whether the mechanism outlined above could produce breaking of CPT (and Lorentz invariance) in our four spacetime dimensions. It would seem natural for this to occur, since there is no apparent reason why four dimensions should be preferentially selected in the higher-dimensional theory. However, no CPT violation has been experimentally detected, so any such breaking must be highly suppressed in the standard model. In a realistic string theory and treating the standard model as an effective low-energy model, the natual dimensionless suppression factor that appears would be the ratio r of the low-energy scale to the Planck scale, $r \sim 10^{-17}$. This suppression factor would produce only a few potentially observable CPT-violating effects, among which are ones in principle detectable in the kaon and other neutral-meson systems [9, 11].

A generic CPT-violating contribution to the effective four-dimensional low-energy theory (the standard model) that could emerge from a compactified string theory could have the form [10, 11]:

$$\mathcal{L} \sim \frac{\lambda}{M^k} \langle T \rangle \cdot \overline{\psi} \Gamma (i\partial)^k \chi + h.c. \quad . \tag{1}$$

Here, $\langle T \rangle$ is the expectation value of a Lorentz tensor T. The four-dimensional fermions ψ and χ are contracted in spinor space through a gamma-matrix structure Γ, with couplings to T possibly involving derivatives $i\partial$. The factors of the (Planck or compactification) mass M must be present on dimensional grounds, and λ is taken to be a dimensionless coupling constant.

A particularly interesting CPT-violating extension of the standard model can be obtained from terms of the form (1) by identifying the fermions ψ and χ with ones appearing in the standard model and requiring that the usual SU(3) × SU(2) × U(1) gauge invariance is maintained. The possible terms compatible with naive power-counting renormalizability have been explicitly given in ref. [12], along with a framework for treating theoretically the accompanying CPT and Lorentz breaking.

The next sections summarize some of the observable consequences of such terms in neutral-meson systems. Other effects are also possible. For example, under suitable circumstances terms of the form (1) could produce baryogenesis in thermal equilibrium [13]. This mechanism for generating the observed baryon asymmetry is distinct from more conventional ones that require nonequilibrium processes and C- and CP-breaking interactions [22].

NEUTRAL-MESON OSCILLATIONS

To investigate possible CPT-violating signals in neutral-meson systems, ψ and χ can be taken as the quarks comprising the neutral meson, denoted generically by P ($P \equiv K$, D, B_d, or B_s). Terms of the form (1) then produce contributions to the 2×2 effective hamiltonian Λ governing the time evolution of the meson system. Within the context of conventional quantum mechanics, there are two kinds of (indirect) CP violation that can appear in Λ: T-violating contributions that preserve CPT, and CPT-violating contributions that preserve T. The corresponding complex parameters are denoted ϵ_P and δ_P, respectively. A plausible theoretical framework for understanding the appearance of a nonzero value of the T-violating parameter ϵ_P exists in the context of the standard model, using the CKM matrix.

The CPT-breaking extension of the standard model mentioned in the previous section provides a basis for understanding the origin of a possible nonzero value of the CPT-violating quantity δ_P in terms of spontaneous CPT and Lorentz breaking as might occur in the string scenario, for example. An analysis shows that δ_P can be expressed within this framework as [10, 11]

$$\delta_P = i \frac{h_{q_1} - h_{q_2}}{\sqrt{\Delta m^2 + \Delta \gamma^2 / 4}} e^{i\hat{\phi}} \quad . \tag{2}$$

In this equation, Δm and $\Delta\gamma$ are mass and rate differences and $\hat{\phi} = \tan^{-1}(2\Delta m/\Delta\gamma)$. These are experimental observables. The quantities $h_{q_j} = r_{q_j}\lambda_{q_j}\langle T\rangle$ originate from the terms (1) and from the effects of the quark-gluon sea, parametrized by r_{q_j}.

Since the underlying fundamental theory is assumed to be hermitian and since the CPT and Lorentz breaking are spontaneous, the quantities h_{q_j} are real. This implies the relationship

$$\text{Im}\,\delta_P = \pm\cot\hat{\phi}\,\text{Re}\,\delta_P \quad , \tag{3}$$

connecting the real and imaginary parts of δ_P through an experimental observable. The small size of the suppression ratio r precludes experimental detection of any direct CPT violation in the decay amplitudes of the P meson, so if CPT violation is indeed detected using neutral mesons then the result (3) would be the primary signature.

EXPERIMENTAL TESTS

Experimental tests of CPT violation can be envisaged in any of the K, D, B_d, and B_s neutral-meson systems [11]. Within the string-based framework described in the previous section, the CPT-violating parameters δ_P given by (2) depend on dimensionless coupling constants λ_{q_j} that are presumably of different magnitude for different quark flavors q_j. This means the δ_P should differ for distinct P mesons. An analogous situation occurs for the standard-model Yukawa couplings, which range over some six orders of magnitude.

One implication of this degree of freedom is that CPT symmetry should be tested experimentally in more than one neutral-meson system. Another is that some startling possibilities might occur in the behavior of heavy neutral mesons. In the B_d system, for instance, there are currently no bounds on CPT violation and the bounds on T violation are relatively weak. It is therefore conceivable that CPT violation could exceed the expected conventional T violation, which would produce unexpected signals in the proposed B factories.

Experiments investigating indirect CP violation use either uncorrelated neutral mesons P or correlated P-$\overline{P}$ pairs arising from quarkonium decays. Typical experimental signatures for CPT and T violation involve asymmetries of decay probabilities into different final states. Appropriate asymmetries with and without time dependence and for both correlated and uncorrelated cases are presently available for all neutral-meson systems. These have been used both for relatively simple theoretical estimates and as input for detailed Monte-Carlo simulations of realistic experimental data, including background effects and acceptances.

In the remainder of this section, I provide a few remarks about the current status of CPT violation in the various neutral-meson systems. The reader is referred to the original literature [9]-[16] for a more complete treatment.

The K system presently offers the only neutral-meson limit on CPT violation. The published bounds [8, 23, 24] correspond to limits on $|\delta_K|$ of order 10^{-3}. Data from various experiments recently completed (e.g., CPLEAR at CERN) or now underway (e.g., KTeV at Fermilab) are likely to lead to an improved bound within the near future.

In the D system, no mixing has yet been observed experimentally and strong dispersive effects make theoretical calculations uncertain. Estimating the CPT reach of future experiments is therefore relatively difficult. Nonetheless, under theoretically favorable circumstances there are some interesting possibilities for placing bounds on δ_D with available techniques and perhaps even from existing data.

The B_d system is of especial interest for CPT tests because it involves the heaviest quark and so might generate the largest CPT violation. Currently, no limit on δ_{B_d} has been published. However, enough data have been obtained to place a bound on δ_{B_d}. A conservative Monte-Carlo simulation with realistic experimental data [15] suggests a limit of order 10% on δ_{B_d} could be extracted by analyzing existing data from CERN and Cornell. In any event, the planned B factories are expected to improve this significantly.

ACKNOWLEDGMENTS

My thanks to Orfeu Bertolami, Don Colladay, Rob Potting, Stuart Samuel, and Rick Van Kooten for pleasant collaborations leading to results presented in this talk. This work was supported in part by the United States Department of Energy under grant number DE-FG02-91ER40661.

REFERENCES

1. J. Schwinger, Phys. Rev. **82** (1951) 914.

2. G. Lüders, Det. Kong. Danske Videnskabernes Selskab Mat.-fysiske Meddelelser **28**, no. 5 (1954).

3. J.S. Bell, Birmingham University thesis (1954); Proc. Roy. Soc. (London) **A 231** (1955) 479.

4. W. Pauli, p. 30 in W. Pauli, ed., *Niels Bohr and the Development of Physics*, McGraw-Hill, New York, 1955l

5. G. Lüders and B. Zumino, Phys. Rev. **106** (1957) 385.

6. R.F. Streater and A.S. Wightman, *PCT, Spin and Statistics, and All That*, Benjamin Cummings, Reading, 1964.

7. R. Jost, *The General Theory of Quantized Fields* (AMS, Providence, 1965).

8. See, for example, R.M. Barnett *et al.*, Review of Particle Properties, Phys. Rev. D **54** (1996) 1.

9. V.A. Kostelecký and R. Potting, Nucl. Phys. B **359** (1991) 545; Phys. Lett. B **381** (1996) 389.

10. V.A. Kostelecký, R. Potting, and S. Samuel, in S. Hegarty et al., eds., *Proceedings of the 1991 Joint International Lepton-Photon Symposium and Europhysics Conference on High Energy Physics*, World Scientific, Singapore, 1992;
V.A. Kostelecký and R. Potting, in D.B. Cline, ed., *Gamma Ray–Neutrino Cosmology and Planck Scale Physics* (World Scientific, Singapore, 1993) (hep-th/9211116).

11. V.A. Kostelecký and R. Potting, Phys. Rev. D **51** (1995) 3923.

12. D. Colladay and V.A. Kostelecký, Phys. Rev. D, in press.

13. O. Bertolami, D. Colladay, V.A. Kostelecký, and R. Potting, Phys. Lett. B **395** (1997) 178.

14. D. Colladay and V.A. Kostelecký, Phys. Lett. B **344** (1995) 259.

15. V.A. Kostelecký and R. Van Kooten, Phys. Rev. D **54** (1996) 5585.

16. D. Colladay and V.A. Kostelecký, Phys. Rev. D **52** (1995) 6224.

17. V.A. Kostelecký and S. Samuel, Phys. Rev. D **39** (1989) 683; *ibid.*, **40** (1989) 1886; Phys. Rev. Lett. **63** (1989) 224; *ibid.*, **66** (1991) 1811.

18. S.W. Hawking, Phys. Rev. D **14** (1976) 2460.

19. D. Page, Phys. Rev. Lett. **44** (1980) 301.

20. R.M. Wald, Phys. Rev. D **21** (1980) 2742.

21. J. Ellis, J.L. Lopez, N.E. Mavromatos, and D.V. Nanopoulos, Phys. Rev. D **53** (1996) 3846.

22. A.D. Sakharov, JETP Lett. **5** (1967) 24.

23. L.K. Gibbons et al., Fermilab-Pub-95/392-E (January 1996); B. Schwingenheuer et al., Phys. Rev. Lett. **74** (1995) 4376.

24. R. Carosi et al., Phys. Lett. B **237** (1990) 303.

NON-UNIVERSAL SOFT SUSY BREAKING, SUSY WIMPS
AND DARK MATTER

Pran Nath
Department of Physics, Northeastern University
Boston, Mass. 02115

R. Arnowitt
Center for Theoretical Physics, Dept. of Physics
Texas A & M University, College Station, TX 77843-4242

Abstract

Non-universalities of soft SUSY breaking parameters in the Higgs sector and in the third generation squark sector are discussed. It is found that these non-universalities are strongly coupled at the electro-weak scale. Neutralino dark matter relic density and event rates in neutralino nucleus scattering are analysed with the inclusion of these non-universalities. It is found that there exist interesting signatures for the non-universalities specifically in the neutralino mass range $m_{\tilde{\chi}_1} \leq$ 65 GeV . An analysis is also given of the effect of more accurate determinations of Ωh^2 on SUSY dark matter. Such determinations are expected in the next generation of satellite experiments.

1 Introduction

Most of the previous analyses of dark matter[1,2] in supergravity unification[3] have been carried out in the framework of the universal soft SUSY breaking at the unification scale[3,4]. The SUSY breaking sector of such theories is described by five parameters which are $m_0, m_{1/2}, A_0, B_0, \mu_0$, where m_0 is the universal scalar mass, $m_{1/2}$ is the universal gaugino mass, A_0 is the universal trilinear coupling, B_0 is the universal bilinear coupling, and μ_0 is the Higgs mixing parameter at the unification scale. After radiative breaking of the electro-weak symmetry one can reduce the number of parameters by using the experimental value of the Z boson mass. The reduced set may be characterized by

$$m_0, m_{1/2}, \mu, tan\beta, sign(\mu) \tag{1}$$

Here μ is the value of μ_0 at the electro-weak scale and, $tan\beta = <H_2> / <H_1>$. The theory described by the above SUSY breaking is very predictive. As an example, one can compute masses of the 32 supersymmetric particles in terms of just four parameters leading to a large number of predictions concerning masses and cross sections.

High Energy Physics and Cosmology
Edited by B.N. Kursunoglu *et al.*, Plenum Press, New York, 1997

The result of universality of the soft SUSY breaking parameters was derived using the generational independence of the Kahler potential[3,4]. However, the framework of supergravity unified theories is more general and allows for a generational dependent Kahler potential[5], which in general leads to non-universality of soft SUSY breaking at the GUT scale. On the phenomenological level the non-universalities are constrained rather stringently in certain sectors because of the limits on flavor changing neutral currents (FCNC). In this paper we shall discuss a specific set of non-universalities which are not strongly constrained by FCNC and investigate their low energy effects and their effects on dark matter. These are non-universalities in the Higgs sector[6,2] and in the third generation squark sector[7].

The outline of the paper is as follows: In Sec. 2 we discuss the Higgs sector non-universalities and non-universalities in the third generation squark sector and show that they are strongly coupled at the electro-weak scale[7]. In Sec. 3 we discuss supersymmetric dark matter and the constraints that it is subjected to. In Sec. 4 we investigate the effects of non-universalities on event rates in neutralino-nucleus scattering. In Sec. 5 we discuss the effects of the more accurate determinations of the Hubble parameter expected in the next round of satellite experiments on the allowed range on the relic density and consequently its effects on the neutralino mass range. Conclusions are given in Sec. 6.

2 Non-universal Soft SUSY Breaking

As discussed in the introduction FCNC constraints play an important role in constraining non-universalities in the soft SUSY breaking sector. These constraints are most stringent in the first two generations, but less so for the Higgs sector and for the third generation sector. For this reason we shall focus here on the non-universalities in these latter two sectors. We parametrise the non-universalities in these sectors by δ_{1-4} defined as follows:

$$m^2_{H_1} = m^2_0(1 + \delta_1), \quad m^2_{H_2} = m^2_0(1 + \delta_2) \tag{2}$$

$$m^2_{\tilde{Q}_L} = m^2_0(1 + \delta_3), \quad m^2_{\tilde{U}_R} = m^2_0(1 + \delta_4) \tag{3}$$

where m_0 is the universal scalar mass of the first two generation masses at the GUT scale M_G. Aside from the non-universalities given above we shall assume that there are no other non-universalities present. With the above boundary conditions we find that at the electro-weak scale one has the following mass spectra for the Higgs and the third generation up squark (stop) masses[7].

$$m^2_{H_1} = m^2_o(1 + \delta_1) + m^2_{1/2}g(t) + \frac{3}{5}S_0p \tag{4}$$

$$m^2_{H_2} = m^2_0\Delta_{H_2} + m^2_{1/2}e(t) + A_o m_{1/2}f(t) + m^2_o h(t) - k(t)A^2_o - \frac{3}{5}S_0p \tag{5}$$

In the above the functions e,f,g,h,k are as defined in Ref. 8, and

$$\Delta_{H_2} = \frac{(D_0 - 1)}{2}(\delta_2 + \delta_3 + \delta_4) + \delta_2 \tag{6}$$

Here D_0 defines the top Landau pole, i.e.,

$$y_0 = \frac{y_t}{E(t)D_0}; D_0 = 1 - 6y_t\frac{F(t)}{E(t)} \qquad (7)$$

where $y_t = h_t^2/(4\pi)^2$ and where h_t is the top Yukawa coupling, and E and F are known functions[8]. In Eq. (5) $p \approx 0.0446$ and S_0 is the contribution of the trace anomaly term[9,7], i.e.,

$$S_0 = Tr(Ym^2) = m_{H_2}^2 - m_{H_1}^2 + \sum_{i=1}^{n_g}(m_{\tilde{q}_L}^2 - 2m_{\tilde{u}_R}^2 + m_{\tilde{d}_R}^2 - m_{\tilde{l}_L}^2 + m_{\tilde{e}_R}^2) \quad (8)$$

where all the masses are at the GUT scale and n_g is the number of quark-lepton generations. Non-universalities also affect the third generation up squark (stop) masses. The stop mass^2 matrix is given by

$$\begin{pmatrix} m_{\tilde{t}_L}^2 & -m_t(A_t + \mu ctn\beta) \\ -m_t(A_t + \mu ctn\beta) & m_{\tilde{t}_R}^2 \end{pmatrix} \qquad (9)$$

where

$$m_{\tilde{t}_L}^2 = m_{\tilde{Q}}^2 + m_t^2 + \Delta_{\tilde{Q}} + (\frac{1}{2} - \frac{2}{3}sin^2\theta_W)M_Z^2 cos2\beta - \frac{1}{5}S_0 p \qquad (10)$$

and

$$m_{\tilde{t}_R}^2 = m_{\tilde{U}}^2 + \Delta_{\tilde{U}} + m_t^2 + (\frac{2}{3})sin^2\theta_W M_Z^2 cos2\beta + \frac{4}{5}S_0 p \qquad (11)$$

and where

$$\Delta_{\tilde{Q}} = \frac{(D_0 - 1)}{6}(\delta_2 + \delta_3 + \delta_4) + \delta_3 \qquad (12)$$

$$\Delta_{\tilde{U}} = \frac{(D_0 - 1)}{3}(\delta_2 + \delta_3 + \delta_4) + \delta_4 \qquad (13)$$

In the above $\Delta_{\tilde{U}}$, $\Delta_{\tilde{Q}}$ and the trace anomaly terms contain all the non-universalities while $m_{\tilde{U}}^2$ and $m_{\tilde{Q}}^2$ are as defined in Ref. (8) and are independent of non-universalities. In addition to the Higgs and the stop masses, the quantity that affects low energy physics importantly is the parameter μ^2. Using the condition of the radiative breaking of the electro-weak symmetry one finds the following expression for μ^2 [see Ref. 7]:

$$\mu^2 = m_0^2 C_1 + A_0^2 C_2 + m_{\frac{1}{2}}^2 C_3 + m_{\frac{1}{2}}A_0 C_4 - \frac{1}{2}M_Z^2 + \frac{3}{5}\frac{t^2 + 1}{t^2 - 1}S_0 p \qquad (14)$$

Here

$$C_1 = \frac{1}{t^2 - 1}(1 - \frac{3D_0 - 1}{2}t^2) + \frac{1}{t^2 - 1}(\delta_1 - \delta_2 t^2 - \frac{D_0 - 1}{2}(\delta_2 + \delta_3 + \delta_4)t^2) \quad (15)$$

$$C_2 = -\frac{t^2}{t^2 - 1}k, \; C_3 = -\frac{1}{t^2 - 1}(g - t^2 e), \; C_4 = -\frac{t^2}{t^2 - 1}f \qquad (16)$$

where $t \equiv tan\beta$. One can now identify the region where the non-universalities are enhanced. One possibility for such an enhancement arises when the universal term proportional to m_0^2 is suppressed. For the large tanβ case this occurs when $D_0 \approx 1/3$. In this case the universal term proportional to m_0^2 vanishes and the non-universality effects are enhanced. A similar enhancement of non-universalities also occurs from the A_0 dependence, although in a somewhat different manner. Here the enhancement occurs when the residue of the Landau pole in A_0 vanishes. Now the residue of the Landau pole is given by

$$A_R \approx A_t - 0.6m_{\tilde{g}} \qquad (17)$$

For the case $A_R = 0$ the effects of the Landau pole term are suppressed and the non-universal effects become dominant. A graphical description of this phenomenon is given in the plot of μ vs A_t in Fig. (1). As expected one finds considerable dispersion in the values of μ due to non-universality in proximity to the point where $A_R = 0$. However, away from this region the Landau pole term begins to dominate and the effect of non-universalities become small.

3 SUSY WIMPS and COLD DARK MATTER

In MSSM one postulates that one of the neutralinos is the LSP so that with the assumption of R parity invariance one has a candidate for cold dark matter (CDM). In supergravity grand unification with SU(2)×U(1) broken via

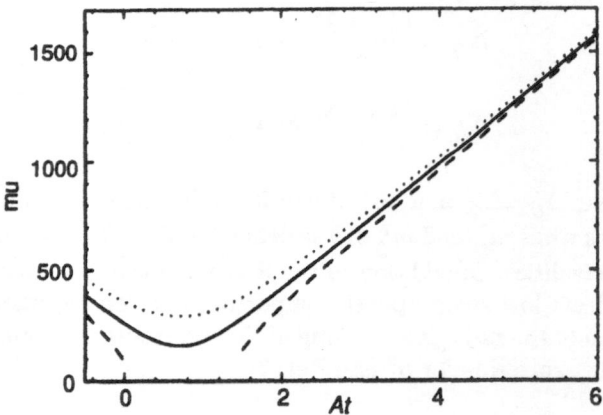

Figure 1. Plot of μ vs A_t when $\delta_3 = 0 = \delta_4$ and (i) $\delta_1 = 0 = \delta_2$(solid), (ii)$\delta_1 = -1 = -\delta_2$(dashed), and (iii)$\delta_1 = 1 = -\delta_2$(dotted). Other parameters are m_0=300 GeV, $m_{\tilde{g}} = 350$ GeV. A_R vanishes when $A_t/m_0 \approx 0.7$. (From Ref. 7.)

radiative effects one deduces that the the lightest neutralino is indeed the LSP over most of the parameter space of the theory. In our analysis we assume R parity invariance and use radiative breaking of the electro-weak symmetry. The parameter space of the model is limited by using the following naturalness constraints

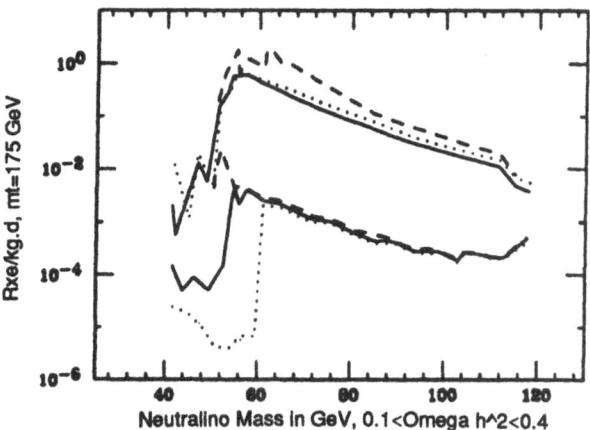

Figure 2. Maximum and minimum of event rates/kg da for xenon for $\mu > 0$ for the case when $\delta_3 = 0 = \delta_4$ and (a)$\delta_1 = 0 = \delta_2$(solid), (b)$\delta_1 = 1 = -\delta_2$(dotted), and (c)$\delta_1 = -1 = -\delta_2$ (dashed) when $0.1 < \Omega_{\tilde{\chi}_1} h^2 < 0.4$, and m_t=175 GeV. (From Ref. [7]).

$$m_0 \leq 1\,TeV,\ m_{\tilde{g}} \leq\ 1TeV,\ tan\beta \leq 25 \tag{18}$$

Using the above range of parameters we compute the relic density of the neutralinos using the standard relation[11]

$$\Omega_{\tilde{\chi}_1^0} h^2 \cong 2.48 \times 10^{-11} \left(\frac{T_{\tilde{\chi}_1^0}}{T_\gamma}\right)^3 \left(\frac{T_\gamma}{2.73}\right)^3 \frac{N_f^{1/2}}{J(x_f)} \tag{19}$$

Here T_γ is the current cosmic microwave background temperature, T_f is the freezeout temperature, N_f is number of massless degrees of freedom at freezeout, $(T_{\tilde{\chi}_1^0}/T_\gamma)^3$ is the reheating factor, $x_f = kT_f/m_{\tilde{\chi}_1}$, and $J(x_f)$ represents the integral over the thermally averaged cross-section $< \sigma v >$ from the current temperature to the freezeout temperature. In computing $< \sigma v >$ we use the accurate method[12,13] which takes into account the integration over the Z and Higgs poles in carrying out the correct thermal average.

The event rate of neutralino nucleus scattering involves contributions from spin dependent interactions from the Z exhange and from squark exchange with chirality diagonal interactions and from the scalar interactions which arise from the exchange of the Higgs and from the L-R chirality terms in the squark exchange.

The total event rate can thus be written as follows[1]:

$$R = [R_{SI} + R_{SD}] \left[\frac{\rho_{\tilde{\chi}_1}}{0.3 GeV\,cm^{-3}}\right] \left[\frac{v_{\tilde{\chi}_1}}{320 km/s}\right] \frac{events}{kg\ da} \tag{20}$$

where $\rho_{\tilde{\chi}_1}$ is the local mass density of $\tilde{\chi}_1$ incident on the detector, and $v_{\tilde{\chi}_1}$ is the incident $\tilde{\chi}_1$ velocity. R_{SI}, is given by

$$R_{SI} = \frac{16m_{\tilde{\chi}_1} M_N^3 M_Z^4}{\left[M_N + m_{\tilde{\chi}_1}\right]^2} |A_{SI}|^2 \tag{21}$$

and the spin dependent rate is given by

$$R_{SD} = \frac{16m_{\tilde{\chi}_1} M_N}{\left[M_N + m_{\tilde{\chi}_1}\right]^2} \lambda^2 J(J+1) |A_{SD}|^2 \tag{22}$$

where J is the nuclear spin and λ is defined by $< N \mid \sum \vec{S}_i \mid N > = \lambda < N \mid \vec{J} \mid N >$. A_{SI} and A_{SD} are the corresponding amplitudes. We note that for large M_N, $R_{SI} \sim M_N$ while $R_{SD} \sim 1/M_N$.

The computation of the event rates is subject to the relic density constraint. To arrive at a reasonable permissible range for $\Omega_{\tilde{\chi}_1} h^2$ we assume an inflationary scenario with $\Omega = 1$, a baryonic component of $\Omega_B = 0.1$, a CDM and a HDM ratio of 2:1, which is consistent with the COBE data, and Hubble constant in the range $0.4 \leq h \leq 0.8$, where h is the Hubble parameter in units of 100 km/s Mpc. With the above assumptions one has

$$0.1 \leq \Omega_{\tilde{\chi}_1} h^2 \leq 0.4 \tag{23}$$

We also subject the analysis to the $b \rightarrow s + \gamma$ constraint. The CLEO value for this branching ratio is[14]

$$BR(b \rightarrow s\gamma) = (2.32 \pm 0.57 \pm 0.35) \times 10^{-4} \tag{24}$$

The Standard Model gives for this branching ratio $(3.28 \pm 0.33) \times 10^{-4}$ for m_t = 174 GeV[15]. In supersymmetry, there are additional diagrams and thus the experimental limits impose important constraints on the parameter space of SUSY models.

4 Event Rates with Universal and Non-universal Soft Breaking

In the analysis below we shall analyse the effects of non-universalities both in the Higgs sector and in the third generation up squark sector on event rates in neutralino nucleus scattering. We focus first on the Higgs sector non-universalities. Here we set $\delta_3 = 0 = \delta_4$ and consider three cases for the values of δ_1 and δ_2: (i)$\delta_1 = 0 = \delta_2$, (ii) $\delta_1 = -1 = -\delta_2$, and (iii) $\delta_1 = 1 = -\delta_2$. The first case is just the universal case, while the other two cases are chosen as two examples of non-universality. The case with universal boundary conditions has been discussed by many authors and recent analyses can be found in Ref.[16]. Here we include this case for comparison. From Eqs. (14) and (15) we see that case (ii) makes a negative contribution to μ^2 while case (iii) makes a positive contribution to μ^2. We shall limit ourselves to the range $|\delta_i| \leq 1 (i=1,2)$. In this range case (ii) with $\delta_1 = -1 = -\delta_2$ gives the largest negative contribution to μ^2 while case (iii) with $\delta_1 = 1 = -\delta_2$ gives the largest positive contribution to μ^2. Thus cases (ii) and (iii) are the extreme limits within the prescribed range of δ_i.

In Fig. 2 the analyses of the maximum and the minimum event rates for xenon for the cases (i)-(iii) listed above are for $\mu > 0$. One notices the dips in the region below $m_{\tilde{\chi}_1} \leq 65$ GeV. These arise because of the rapid annihilation in the vicinity of the Z pole and the Higgs pole. Another interesting feature in the region $m_{\tilde{\chi}_1} \leq 65$ GeV is some characteristic signatures which distinquish the universal and the non-universal cases. Thus a comparison of cases (i) and (ii) shows that the minimum event rates for case (ii) are enhanced over case(i) by a factor of about O(10) because as discussed above μ^2 receives a negative contribution and can turn negative over a part of the parameter

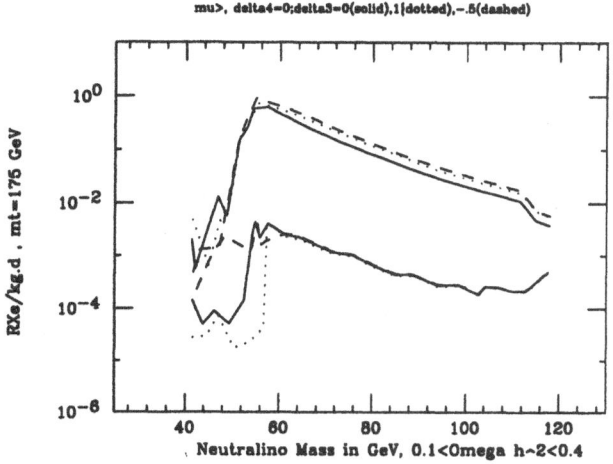

Figure 3. Maximum and minimum of event rates/kg da for xenon for the case when $\delta_1 = \delta_2 = \delta_4 = 0$ and (a)$\delta_3 = 0$(solid), (b)δ_3=1(dotted), and (c)δ_3=-1 (dashed) when $0.1 < \Omega_{\tilde{\chi}_1} h^2 < 0.4$, and m_t=175 GeV. (From Ref. 7.)

space eliminating such points from the allowed range. The eliminated part of the parameter space contains low values of tanβ. Since low values of tanβ are eliminated the minimum event rates tend to increase which is what is seen in Fig. 2. An opposite effect occurs for case(iii). Here μ^2 increases and the minimum event rates decrease further. For values of $m_{\tilde{\chi}_1} \geq 65$ GeV the Landau pole term begins to dominate and the effect of non-universalities is reduced.

In the above we assumed non-universality only in the Higgs sector. However, as was discussed in Sec. 2 there is a strong coupling of the non-universality in the Higgs sector and in the third generation sector. We discuss now the non-universalities in the third generation sector. One can correlate in an approximate manner the effects of the non-universality in the third generation with non-universalities in the Higgs sector by examining Δ_{H_2}. Assuming $M_G = 10^{16.2}$ GeV, $\alpha_G = 1/24$ and $m_t = 175\ GeV$ one has $D_0 \simeq 0.27$, and one may write Δ_{H_2} in the form

$$\Delta_{H_2} \simeq 0.64\delta_2 - 0.36(\delta_3 + \delta_4) \qquad (25)$$

From the above we find that a positive δ_2 in Δ_{H_2} can be simulated by negative values of δ_3 or δ_4, and a reverse situation holds for a negative value of δ_2. The effects of non-universalities in δ_3 are exhibited in Fig. 3 for the case when $\delta_1 = \delta_2 = \delta_4 = 0$. One finds that the general observations made above is

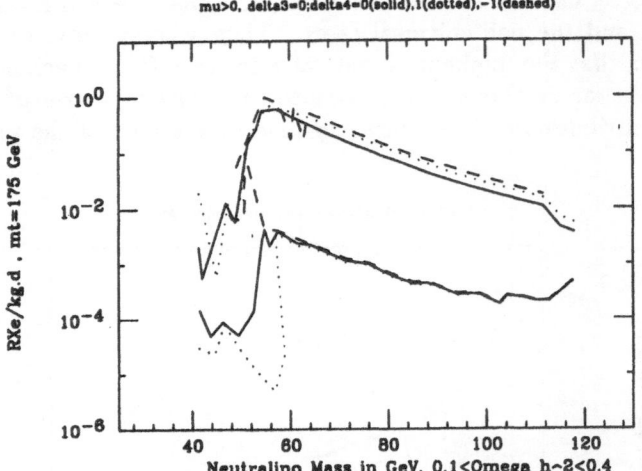

Figure 4. Maximum and minimum of event rates/kg da for xenon for the case when $\delta_1 = \delta_2 = \delta_3 = 0$ and (a)$\delta_4 = 0$(solid), (b)$\delta_4 = 1$(dotted), and (c)$\delta_4 = -1$ (dashed) when $0.1 < \Omega_{\tilde{\chi}_1} h^2 < 0.4$, and $m_t = 175$ GeV. (From Ref. 7.)

borne out by the analysis of the curves displayed. An analysis similar to the above for the case of non-universalities in δ_4 is given in Fig. 4. Again the general observations made above hold in this case as well. We emphasize, however, that the correlations between the effects of non-universalities in the Higgs sector and in the third generation sector are of a qualitative nature only, and the full dynamics for the two cases is quantitatively different.

The analysis given above was for the case $\mu > 0$. Here the $b \to s + \gamma$ constraint is less severe. However, for the $\mu < 0$ case, $b \to s + \gamma$ constraint is far more severe. Here the maximum event rates are a factor of $O(10^{-2} - 10^{-3})$ smaller than for the $\mu > 0$ case.

5 Accurate Ωh^2 Determination

As mentioned in Sec. 1 it is possible that the Hubble parameter may be determined to within $O(10\%)$ or better accuracy in the next round of satellite experiments. An accurate measurement such as this is likely to have a very signifcant effect on the SUGRA parameter space. To study the effect that an accurate measurement can have we consider for illustration the following narrow range of $\Omega_{\tilde{\chi}_1} h^2$

$$0.225 < \Omega_{\tilde{\chi}_1} h^2 < 0.275 \qquad (26)$$

The analysis gives us an upper bound on the neutralino mass of $m_{\tilde{\chi}_1} \approx 90$ GeV. Using the constraint of grand unification we find that the this limit requires that the gluino mass lie in the range $m_{\tilde{g}} < 650$ GeV. An upper bound such as this makes the discovery of the gluino more accessible to an upgraded Tevatron and would certainly be accessible to LHC. Thus we see that more accurate constraints on the cosmological parameters would produce constraints on the sparticle masses which would be testable at supercolliders.

6 Conclusions

In this paper we have investigated the effects of non-universalities in the Higgs sector and in the third generation squark sector and shown that they are strongly coupled at the electro-weak scale. We investigated the effects of these non-universalities on event rates in neutralino-nucleus scattering and found significant effects in the region of neutralino masses less than about 65 GeV. In this region we found that the minimum event rates could be increased or decreased by a factor of O(10). The effect that a more accurate determination of the Hubble parameter may have on constraining the SUGRA parameter space was also investigated. It is found that a narrower range of the Hubble parameter, and thus a narrower range of Ωh^2 will limit severly the allowed mass range of the neutralino, and hence the determinations of the next generation of satellite experiments will have observable consequences at supercolliders. Another important result of the analysis is that the event rates lie in a wide range from $O(1\text{-}10^{-5})$ events/kg da. Thus although data from the dark matter detectors currently in operation has begun to come in[17] and a part of the parameter space of the supergravity models is accessible in these detectors, one needs more sensitive detectors[18] to sample a majority of the parameter space of SUGRA models.

Acknowledgments

This research was supported in part by NSF grant numbers PHY-96020274 and PHY-9411543.

References

1. For a review see, G. Jungman, M. Kamionkowski and K. Greist, Phys. Rep. **267**,195(1995).
2. An exception is the analysis by V. Berezinsky, A. Bottino, J. Ellis, N. Forrengo, G. Mignola, and S. Scopel, Astropart. Phys.5:1-26(1996), where non-universalities of soft SUSY breaking were considered.
3. A.H. Chamseddine, R. Arnowitt and P. Nath, Phys. Rev. Lett. **29**, 970 (1982). For reviews see P.Nath, R. Arnowitt and A.H.Chamseddine, "Applied N = 1 Supergravity" (World Scientific, Singapore, 1984); H.P. Nilles, Phys. Rep. **110**, 1 (1984); R. Arnowitt and P. Nath, Proc. of VII J.A. Swieca Summer School ed. E. Eboli (World Scientific, Singapore, 1994).
4. R. Barbieri, S. Ferrara, and Savoy, Phys. Lett. **B119**,343(1982); L.Hall, J.Lykken and S.Weinberg, Phys. Rev. **D27**,2359(1983); P. Nath, R. Arnowitt and A.H. Chamseddine, Nucl. Phys. **B227**,121(1983).
5. S.K. Soni and H.A. Weldon, Phys. Lett. **B126**, 215(1983);V.S. Kaplunovsky and J. Louis, Phys. Lett.**B306**, 268(1993).
6. D. Matalliotakis and H.P. Nilles, Nucl.Phys.**B435**, 115(1995); M. Olechowski and S. Pokorski, Phys.Lett. **B344**, 201(1995); N. Polonski and A. Pomarol,Phys.Rev.**D51**,6532(1995).

7. P. Nath and R. Arnowitt, hep-ph/9701301.

8. L. Ibanez, C. Lopez, and C. Munos, Nucl. Phys.**B256**, 218(1985).

9. S. Martin and P. Ramond, Phys. Rev. **D48**, 5365(1993).

10. P. Nath, J. Wu and R. Arnowitt, Phys.Rev.**D52**,4169(1995).

11. For a review see E.W. Kolb and M.S. Turner, "The Early Universe" (Addison-Wesley, Redwood City, 1989).

12. K. Greist and D. Seckel, Phys. Rev. **D43**, 3191 (1991); P. Gondolo and G. Gelmini, Nucl. Phys. **B360**, 145 (1991).

13. R. Arnowitt and P. Nath, Phys. Lett. **B299**, 58 (1993); **B303**, 403 (1993) (E); P. Nath and R. Arnowitt, Phys. Rev. Lett. **70**, 3696 (1993); H. Baer and M. Brhlick, Phys.Rev.**D53**,597(1996).

14. M.S. Alam et al. (CLEO Collaboration), Phys. Rev. Lett. **74**, 2885 (1995).

15. K. Chetyrkin, M. Misiak, and M. Munz, hep-ph/9612313.

16. R. Arnowitt and P. Nath, Mod. Phys. Lett. **A 10**,1257(1995); P. Nath and R. Arnowitt, Phys. Rev. Lett.**74**,4592(1995); R. Arnowitt and P. Nath, Phys. Rev. **D54**,2374(1996).

17. R. Bernabei,et.al., Phys. Lett. **B389**, 757(1996); P. F. Smith et. al., Phys. Lett. **379**, 299(1996).

18. D. Cline."On a Discriminatory Liquid Xenon SUSY Wimp Detector", Nucl. Phys. B (in press); P. Benetti et al, Nucl. Inst. and Method for Particle Physics Research, **A307**,203 (1993).

SEEING PLANCK SCALE PHYSICS AT ACCELERATORS

R.Arnowitt[1] and Pran Nath[2]

[1]Center for Theoretical Physics, Department of Physics
Texas A&M University, College Station, TX 77843-4242
[2]Department of Physics, Northeastern University
Boston, MA 02115

Abstract

Much current theoretical analysis is based on the hypothesis that the physics beyond the Standard Model is a consequence of new principles that occur at the Planck scale. The question arises whether such principles can ever be directly tested. We show here that for a significant class of models, hypotheses at the string or Planck scale can indeed be directly tested to relatively high precision by linear colliders. Three classes of models are examined: those with universal SUSY soft breaking at the string scale; those with a horizontal symmetry at the string or Planck scale, and simple Calabi-Yau superstring models with dilaton and moduli SUSY breaking.

INTRODUCTION

While the Standard Model is a remarkably successful theory, having been subjected to numerous high precision experimental tests, there are many aspects about it that are not understood, e.g. Yukawa couplings, CKM parameters, etc. There has been much theoretical analysis based on the assumption that these items are a consequence of new physical principles arising at or near the Planck scale, $M_{Pl} \cong 2.4 \times 10^{18}$ GeV. The strongest direct evidence that the ultra high energy domain may influence low energy phenomena resides in the success of supersymmetric (SUSY) grand unification, i.e. that the three coupling constants α_1, α_2, α_3 unify at a GUT scale $M_G \simeq 2 \times 10^{16}$ GeV to a value $\alpha_G \simeq 1/24$ for a minimal SUSY particle spectrum with one pair of Higgs doublets. First observed in the 1990 LEP data [1], this result has stood the test of time, both with refinements in the data and refinements in the theoretical treatment [including for the latter SUSY threshold effects at $M_S \simeq 100$ GeV -1 TeV, GUT scale threshold effects near M_G, and possible small Planck physics effects from non-renormalizable operators (NRO)].

We will use supergravity (SUGRA) grand unification [2] to analyse these questions here. There has been much discussion over the past year as to at what scale SUSY breaks in such models. We will assume here that supersymmetry breaks in a hidden sector at a scale above M_G, e.g. at $O(M_{Pl})$. These models have a number of undetermined aspects presumably representing Planck scale physics, and we will examine three types of post-GUT assumptions that might be made:

1. Models with universal SUSY soft breaking masses at the string scale $M_{str} \simeq 5 \times 10^{17}$ GeV.

2. Models with $SU(2)_H$ horizontal symmetry.

3. String models with Calabi-Yau compactification.

We will see in the above examples that linear colliders have in fact the ability to probe physics in the post-GUT region up to the string or Planck scales to a remarkable degree of precision, distinguish between different models and actually measure parameters that are expected to be predictions of string models.

SUPERGRAVITY MODELS

We review briefly the basic elements of supergravity models. These models depend in general on three functions of the scalar fields $\{\phi_i\}$ (squarks, sleptons etc.): $f_{\alpha\beta}(\phi_i)$, the gauge kinetic function, $K(\phi_i, \phi_i^\dagger)$, the Kahler potential and $W(\phi_i)$, the superpotential. $f_{\alpha\beta}$ modifies the gauge and gaugino kinetic energies (e.g. $f_{\alpha\beta} F_{\mu\nu}^\alpha F^{\mu\nu\beta}$, α, β = gauge indices). K enters into the scalar and chiral partner kinetic energies (e.g. $K_j^i \partial_\mu \phi_i \partial^\mu \phi_j^\dagger$ where $K_j^i \equiv \partial^2 K (\partial\phi_i \partial\phi_j^\dagger)$) and elsewhere. W and K enter in the Lagrangian only in the combination

$$G(\phi_i, \phi_i^\dagger) = \kappa^2 K(\phi_i, \phi_i^\dagger) + \ell n[\kappa^6 \mid W(\phi_i) \mid^2] \tag{1}$$

where $\kappa \equiv 1/M_{P\ell}$. To maintain the gauge hierarchy we assume the superpotential decomposes into a "physical" and a "hidden" sector,

$$W(\phi_i) = W_{phys}(\phi_a) + W_{hid}(z) \tag{2}$$

Here $\{\phi_i\} = \{\phi_a, z\}$ where ϕ_a are physical fields and z are fields whose VEVs spontaneously break supersymmetry and obey $\langle z \rangle = O(M_{P\ell})$ and $\kappa^2 \langle W_{hid} \rangle = O(M_S)$.

The mass dimensions of the basic functions are $[f_{\alpha\beta}] = (\text{mass})^0$, $[K] = (\text{mass})^2$, and $[W] = (\text{mass})^3$. It is convenient to introduce the dimensionless variables $x \equiv \kappa z$ with $\langle x \rangle = O(1)$ and expand $f_{\alpha\beta}$, K and W in powers of ϕ_a with higher terms scaled by κ. Thus one can write

$$f_{\alpha\beta}(\phi_i) = c_{\alpha\beta}(x) + \kappa c_{\alpha\beta}^a(x)\phi_a + \cdots \tag{3}$$

$$\begin{aligned} K(\phi_i, \phi_i^\dagger) &= \kappa^{-2} c(x, x^\dagger) + c_b^a \phi_a \phi_b^\dagger \\ &+ (c^{ab}\phi_a\phi_b + c_{ab}\phi_a^\dagger\phi_b^\dagger) + \kappa c_{bc}^a \phi_a\phi_b\phi_c^\dagger\phi_d^\dagger + \cdots \end{aligned} \tag{4}$$

$$W_{phys}(\phi_i) = \frac{1}{6}\lambda^{abc}\phi_a\phi_b\phi_c + \frac{1}{24}\kappa\lambda^{abcd}\phi_a\phi_b\phi_c\phi_d + \cdots \tag{5}$$

The assumption that after SUSY breaking, the VEVs of $c_{\alpha\beta}$, c, c_b^a etc. are all of $O(1)$, implies that the higher order terms scaled by κ are of $O(1/M_{P\ell})$ and presumably represent Planck scale physics corrections (e.g. arising in string theory from integrating out the towers of Planck mass states). The terms with dimensionless coupling constants are accessible to low energy physics (e.g. λ^{abc} are the Yukawa couplings). The holomorphic terms in K ($c^{ab}\phi_a\phi_b$ etc.) can be transferred to W_{phys} by a Kahler transformation, and then give rise naturally to a μ term in W_{phys} after SUSY breaking of size $\mu = O(M_S)$ [3].

The spontaneous breaking of supersymmetry gives rise to the SUSY soft breaking masses [2,4]. For SUSY breaking above M_G, the pattern of soft breaking masses must obey the symmetries of the GUT group $\mathcal{G}$. We consider here the cases where $\mathcal{G}$ contains an SU(5) subgroup (e.g. SU(N), $N \geq 5$; SO(N), $N \geq 10$; E_6) and label the light matter at M_G by their SU(5) quantum numbers. Thus for three generations of 10 and $\bar{5}$ representations labeled by a = 1,2,3 and one pair of Higgs ($\mathcal{H}_1 = \bar{5}$, $\mathcal{H}_2 = 5$) one has

$$
\begin{aligned}
10_a &= \{q_a = (\tilde{u}_{La}, \tilde{d}_{La}); \ u_a \equiv \tilde{u}_{Ra}; \ e_a \equiv \tilde{e}_{Ra}\} \\
\bar{5}_a &= \{\ell_a \equiv (\tilde{\nu}_{La}, \tilde{e}_{La}); \ d_a \equiv \tilde{d}_{Ra}\}
\end{aligned} \tag{6}
$$

(For the flipped SU(5) model [5], one interchanges $\tilde{u}$ and $\tilde{d}$, $\tilde{\nu}$ and $\tilde{e}$ with $\tilde{e}_R$ appearing in an extra SU(5) singlet.) Each representation can have an independent soft breaking mass which we parameterize as

$$
\begin{aligned}
m_{10_a}^2 &= m_o^2(1 + \delta_a^{10}); \quad m_{\bar{5}_a}^2 = m_o^2(1 + \delta_a^{\bar{5}}) \\
m_{H_1}^2 &= m_o^2(1 + \delta_1); \quad m_{H_2}^2 = m_o^2(1 + \delta_2)
\end{aligned} \tag{7}
$$

where $m_o = O(M_S)$. Thus a general model of this type can have eight soft breaking scalar masses, though specific models may have fewer parameters. (In addition there are, of course, gaugino masses and cubic and quadratic soft breaking parameters.) The δ_i measure the amount non-universality in the scalar soft breaking masses. One has at the grand unification scale M_G the 15 relations:

$$
\delta_{qa} = \delta_{ua} = \delta_{ea} = \delta_a^{10}; \quad \delta_{\ell a} = \delta_{da} = \delta_a^{\bar{5}} \tag{8}
$$

Below M_G, the standard model gauge group holds for many models. One may make contact with low energy physics by running the renormalization group equation (RGE) from M_G down to the electroweak scale M_Z. Remarkably, the spontaneous breaking of supersymmetry at M_G triggers the breaking of SU(2) x U(1) at low energy [6,7], the scale at which electroweak breaking occurs being determined in large part by the top quark mass, which we take here to be $m_t = 175$ GeV.

Similarly, one may try to make contact with Planck scale physics by running the RGE upwards to scales above M_G. Here, however, results depend upon the particle spectrum and gauge group $\mathcal{G}$ above M_G. We will see that linear colliders are sensitive to both these types of model dependences and hence will be able to distinguish between different possibilities.

STRING SCALE UNIVERSALITY

Non-universal soft breaking masses at M_G can arise from running the RGE down from a higher scale, the non-universal effects being due to different Yukawa couplings etc. We consider in this section the case where all soft breaking masses are universal at the string scale M_{str} [8]. Above M_G, the gauge group $\mathcal{G}$ is unbroken and different gauge groups will give different results.

(i) $\mathcal{G} = SU(5)$

We restrict the discussion to third generation effects, which have the largest Yukawa couplings, and assume that above M_G there is a $10 + \bar{5}$ of matter and a $5 + \bar{5} + 24$ of Higgs representations present. The superpotential has the form

$$
\begin{aligned}
W_{phys} &= [h_t(10)(10)H_5 + h_b(10)(\bar{5})H_{\bar{5}}] + [Mtr(24)^2 + \lambda_1 tr(24)^3 \\
&+ \lambda_2 H_5(24)H_{\bar{5}} + \mu_o H_5 H_{\bar{5}}]
\end{aligned} \tag{9}
$$

where $M=O(M_G)$. One may chose the reference mass m_o to be m_{10}, and then one has the following for the remaining scalar soft breaking masses:

$$m_5^2 = m_o^2(1 + \delta_5); \quad m_{H_{1,2}}^2 = m_o^2(1 + \delta_{1,2}) \tag{10}$$

The values of δ_5, $\delta_{1,2}$ at M_G are model dependent, depending on the coupling constants and interactions chosen in Eq. (9). Characteristically one finds from running the RGE from M_{str} to M_G that $|\delta_i| \lesssim 1/2$. We will in the following assume

$$-1 \leq \delta_i \leq 1 \tag{11}$$

The values of m_o, δ_5, $\delta_{1,2}$ are not determined by supergravity theory, but presumably will be set by future Planck scale physics. However, they can be experimentally measured by using the RGE and relating them to electroweak scale quantities. Thus one has [7,9]

$$m_o^2 = m_{\tilde{e}_R}^2 - 0.151 m_{1/2}^2 - sin^2\theta_W M_Z^2 cos2\beta \tag{12}$$

$$m_o^2\delta_5 = m_{\tilde{e}_L}^2 - m_{\tilde{e}_R}^2 - 0.377 m_{1/2}^2 + \left(\frac{1}{2} - sin^2\theta_W\right) M_Z^2 cos2\beta \tag{13}$$

where $\tilde{e}_{R,L}$ are the R,L selectrons, $m_{1/2} = (\alpha_G/\alpha_2)\tilde{m}_2$ where $\tilde{m}_2$ is the SU(2) gaugino mass, and $tan\beta = \langle H_2 \rangle / \langle H_1 \rangle$. The numerics in the above formulae arise from running the RGE from M_G to the electroweak scale. Studies have been made as to how accurately these parameters can be measured at the proposed Next Linear Collider (NLC) [10-13]. Thus it is expected that $m_{\tilde{e}_{R,L}}$ could be measured to 1% accuracy, $\tilde{m}_2$ to 3%, $tan\beta$ to 10% and α_G perhaps to 3%. Suppose, for example, actual measurements at the NLC found $m_{\tilde{e}_L} = 240$ GeV, $\tilde{m}_2 = 120$ GeV and $tan\beta = 5$ with the above accuracies. Then Eqs. (12,13) imply [9]

$$m_o = (187 \pm 3)GeV; \quad \delta_5 = 0.206 \pm 0.031 \tag{14}$$

The above discussion shows that the value of m_o and the existance of non-universal soft breaking (i.e. $\delta_5 \neq 0$) can be established at the NLC to remarkable accuracy. Further, there are many other ways of measuring m_o and δ_5, e.g via squark masses which would act as a check on the validity of the model, and allow one to reduce the experimental errors.

The parameters δ_1 and δ_2 could be determined from μ and m_A, where A is the CP odd neutral Higgs boson. One has [7,9]

$$
\begin{aligned}
\mu^2(t^2 - 1) &= [\delta_1 - \frac{1}{2}t^2(1 + D_0)\delta_2]m_0^2 + [1 - \frac{1}{2}t^2(3D_0 - 1)]m_0^2 \\
&+ [0.528 + t^2(3.22 - 3.80D_0 + 0.060D_0^2)]m_{1/2}^2 \\
&+ \frac{1}{2}t^2(1 - D_0)\frac{A_R^2}{D_0} - \frac{1}{2}M_Z^2(t^2 - 1)
\end{aligned} \tag{15}
$$

$$
\begin{aligned}
m_A^2 \frac{t^2 - 1}{t^2 + 1} &= \left[\delta_1 - \frac{1}{2}(1 + D_0)\delta_2 + \frac{3}{2}(1 - D_0)\right]m_0^2 \\
&+ [3.22 - 3.80D_0 + 0.060D_0^2]m_{1/2}^2 + \frac{1}{2}(1 - D_0)\frac{A_R^2}{D_0} \\
&- \frac{t^2 - 1}{t^2 + 1}M_Z^2
\end{aligned} \tag{16}
$$

where $D_0 \cong 1-(m_t/200 \sin\beta)^2$, $A_R \cong A_t - 0.613 m_{\tilde{g}}$, $\tilde{g}$ is the gluino and A_t is the t-quark cubic soft breaking parameter at the electroweak scale. ($D_0 = 0$ is the t-quark Landau pole and A_R is the residue at the pole.) For a linear collider (LC) of sufficient energy that heavy neutralinos ($\tilde{\chi}_{3,4}$) can be produced [14] and A pair produced [15], both μ and m_A can be determined to about 2%. One expects A_R would have an error of about 5%. For example, if measurements were to yield $m_o = 200$ GeV, $\mu = 325$ GeV, $m_A = 400$ GeV, one finds from Eqs. (15,16) that the GUT scale parameters are

$$m_{H_1} = (256 \pm 15)GeV; \quad m_{H_2} = (144 \pm 35)GeV \tag{17}$$

and hence $\delta_1 = 0.634 \pm 0.220$ and $\delta_2 = 0.485 \pm 0.178$, allowing a test of whether the Higgs masses are universal at M_G.

Further, from Eq. (8) there are three differences where this model predicts that the non-universal soft breaking effects should cancel out:

$$m_{\tilde{u}_L}^2 - m_{\tilde{u}_R}^2; \quad m_{\tilde{u}_L}^2 - m_{\tilde{e}_R}^2; \quad m_{\tilde{e}_L}^2 - m_{\tilde{d}_R}^2 \tag{18}$$

Finally we note that if the model is correct, one can use the RGE and run all the scalar soft breaking masses up to M_{str} where they all should approach a common value $(m_0)_{str}$. Thus one can even determine in this way the value of M_{str} experimentally, i.e. it would be the scale at which the masses unify.

(ii) $\mathcal{G} = SO(10)$

One can carry out a similar analysis for the SO(10) group. We consider here the case where SO(10) breaks directly to the Standard Model at M_G and the 5+$\bar{5}$ of SU(5) Higgs both reside in the same 10 of SO(10). Under these circumstances, all the SU(5) relations considered above still hold and in addition there is the constraint [16] $\delta_5 = \delta_2 - \delta_1$. For the parameters discussed above one finds

$$\delta_5/(\delta_2 - \delta_1) = 0.160 \pm 0.040 \tag{19}$$

which would imply for this case that the SO(10) relation is strongly violated. Thus one could experimentally distinguish between different gauge groups.

(iii) Distinguishing Between Different Post-GUT Groups

Much of the physics below the GUT scale is insensitive to the nature of the GUT group that holds above M_G. However, the above analysis shows that linear colliders should be able to give information about physics beyond M_G, and distinguish between different GUT groups. Thus if, experimentally one finds that $\delta_{u_a} \neq \delta_{e_a}$ and δ_{u_a}, $\delta_{e_a} \neq 0$, and if $\delta_{d_a} \neq \delta_{\ell a}$, then the SM gauge group would be valid, and SU(5) or SO(10) would be eliminated. If, however, one finds $\delta_{u_a} = \delta_{e_a} = 0$ and $\delta_{d_a} = \delta_{\ell a}$ then either SU(5) or SO(10) could be valid above M_G, and one could distinguish between them by checking whether the relation $\delta_5 = \delta_2 - \delta_1$ holds. The size of the soft breaking parameters, and generation dependence could give additional information about the post-GUT physics.

HORIZONTAL SYMMETRIES

One of the important and unresolved problems in the SM is the hierarchy of quark and lepton masses. This is related to the problem of suppressing flavor changing neutral currents (FCNC) arising at the loop level from both quarks and squark interactions.

An interesting approach to these questions involves imposing an $SU(2)_H$ horizontal symmetry in generation space [18]. Here one puts the first two generations into an $SU(2)_H$ doublet, and the third generation into an $SU(2)_H$ singlet. We consider here the case where the GUT group is chosen to be [9] $\mathcal{G} = SU(5) \times SU(2)_H$.

In such models, one assumes that $SU(2)_H$ is broken by $SU(2)_H$ doublet Higgs fields ϕ_i whose VEV splits the quark and lepton masses in the first two generations. In order to get the experimental pattern of quark and lepton masses, one requires [18] $\epsilon = \langle \phi_i \rangle / M_{Pl} \approx 1/10$. The picture one has, then, is that supersymmetry breaks at Planck scale, $SU(2)_H$ at the string scale ($M_{str} \simeq M_{Pl}/10$) and finally $SU(5)$ at the GUT scale ($M_G \simeq M_{str}/10$), and we will assume this in the following.

The breaking of $SU(2)_H$ produces an $O(\epsilon^2)$ splitting in the first two squark and slepton generations. This is small and thus helps to suppress FCNC in the K^o - $\bar{K}^o$ and $K_L \rightarrow \mu^+ \mu^-$ SUSY box diagrams. Thus neglecting this $O(1\%)$ effect, one has at M_G the following pattern of scalar soft breaking masses: $(m_{i\bar{5}})^2 = m_o^2 (1 + \delta_{\bar{5}}^d)$; $m_{i10}^2 = m_o^2 (1 + \delta_{10}^s)$; $m_{\bar{5}}^2 = m_o^2 (1 + \delta_{\bar{5}}^s)$; $m_{H_{1,2}}^2 = m_o^2 (1 + \delta_{1,2})$ where $i = 1,2$ is a generation index, and masses without this label are singlet third generation masses. In the above, we have taken the first two generation masses of the 10 representation as the reference mass i.e. $m_o = m_{i10}$. The above model thus depends on six mass parameters.

While the small splitting of the doublet degeneracy will be difficult to measure directly, much of the other structure will be accessible to a LC. There are eight independent sfermion mass measurements which can be used to determine the singlet parameter δ_{10}^s and four that can be used to determine the doublet singlet splitting $\delta_{\bar{5}}^d - \delta_{\bar{5}}^s$. Thus, if each measurement is accurate to 15%, and the model were valid, one could determine each parameter to (5-10)% accuracy, giving a reasonable test of the model. Further, unlike models which assume universality at M_{str}, using the RGE to proceed to scales above M_G would not be expected to lead to the doublet and singlet soft breaking masses unifying at the higher string scale, since the third generation is in a different $SU(2)_H$ representation from the first two. Thus this model is distinguishable from those of the previous section.

SUPERSTRING MODELS

The mechanism for supersymmetry breaking in superstring theory is not understood at present, and as a consequence one cannot make phenomenological predictions in string theory from first principles. However, it has been suggested that SUSY breaking in string theory may arise from VEV formation of the dilaton field and the T and U moduli fields [19]. We consider here some simple Calabi-Yau models where $\mathcal{G}_j = E_6 \times E_8$ with only a single T modulus [20]. For these cases, the soft breaking masses arising from dilaton and T moduli VEVs are actually universal at M_{str}. However, the string theory imposes additional constraints which can be experimentally tested at linear colliders. Thus one has the following relations at M_{str} [21].

$$m_{1/2} = \sqrt{3} \sin \theta e^{-i\gamma_s} m_{3/2} \tag{20}$$

$$m_o^2 = [\sin^2 \theta + \cos^2 \theta \Delta(T, T^*)] m_{3/2}^2 \tag{21}$$

$$A_o = -\sqrt{3} [\sin \theta e^{-i\gamma_s} + \cos \theta e^{-i\gamma_T} \omega(T, T^*)] m_{3/2} \tag{22}$$

where $m_{3/2}$ is the gravitino mass, $m_{1/2}$ is the universal gaugino mass, A_o is the universal cubic soft breaking parameter, $\gamma_{S,T}$ are possible PC violating phases, and θ is the dilaton-Goldstino angle. Δ and ω include σ model corrections and instanton corrections to the Kahler potential. For simplicity we set $\gamma_{S,T}$ to zero. Specific string models determine θ, Δ and ω. We leave these arbitrary for the moment.

From Eqs. (22) and (23) one has

$$\frac{m_o^2}{m_{1/2}^2} = \frac{1}{3}[1 + \Delta ctn^2\theta] \qquad (23)$$

From the discussion above, one saw that m_o could be determined at a LC with error of about 2% and $m_{1/2}$ with error of about 5%. Thus using the parameters of the previous analyses ($\tilde{m}_2 = 120$ GeV, $m_o = 187$ GeV) one finds

$$\Delta ctn^2\theta = 3.73 \pm 0.25 \qquad (24)$$

A_o can be related to low energy parameters by the RGE yielding the relation $A_o = A_R/D_o$ - 2.20 $m_{1/2}$, and for example chosing $A_t = 285$ GeV, $tan\beta = 5$ one finds $A_o/m_{1/2} = -1.539 \pm 0.047$ which yields using Eqs. (20,22):

$$\omega \ ctn \ \theta = 0.539 \pm 0.047 \qquad (25)$$

Specific Calabi-Yau compactifications determine Δ and ω. Thus Eqs. (24,25) allow for two experimental determinations of θ. We consider two models.

The value of Δ and ω can be calculated in the large Calabi-Yau radius limit [22]. Thus for ReT = 5, the one modulus models [20] give average values of [21] $\Delta \cong 0.40$, $\omega \cong 0.17$. Then Eqs. (24) and (25) yield $\mid ctn \ \theta \mid = 3.05 \pm 0.14$; $ctn \ \theta = 3.17 \pm 0.28$. We see for the above parameters, that these two values are consistent. One can then determine the gravitino mass from (22) yielding $m_{3/2} = (276\pm 18)$ GeV. The model can then be subject to other experimental tests for universality at the string scale, as described above. Both θ and $m_{3/2}$ are aspects of string supersymmetry breaking. Thus it is possible to experimentally verify at a LC what these predictions would be for this model, once an understanding of supersymetry breaking in string theory is obtained.

As a second model one choses the value Im T = 1/4 which maximizes Δ. Then Re T = 5 yields [21] $\Delta = 1.62$, $\mid \omega \mid = 0.64$. Eqs. (24) and (25) would then yield for the above parameters $\mid ctn\theta \mid = 1.516 \pm 0.071$ and $\mid ctn\theta \mid = 0.842 \pm 0.073$ showing that it would be experimentally possible to rule out this string model.

CONCLUSIONS

LEP has allowed for precision tests of the Standard Model, i.e. physics $\lesssim 100$ GeV, but also from grand unification analyses, it has been possible to probe physics up to the GUT scale. Linear colliders and the LHC will be able to unravel the new physics that lies in the TeV region above the electroweak scale. In addition, we have seen here that linear colliders will be able to probe physics up to the Planck scale and test assumptions made in the post-GUT domain. We have illustrated this here with three classes of models:

- Supergravity models with universal soft breaking at M_{str}. The predicted loss of unversality at M_G could be well measured, different gauge groups distinguished [e.g. SU$'$5), SO(10)], and the value of M_{str} measured.

- Models with horizontal symmetry, e.g. SU(2)$_H$. The general SU(2)$_H$ symmetry is easily observable, and the soft breaking parameters of such models well measured.

- Simple Calabi-Yau string models. Different compactifications could be distinguished and explicit string quantities (e.g. nature of goldstino, value of the gravitino mass $m_{3/2}$) can be well measured.

Thus linear colliders are potentially very powerful experimental tools for unraveling physics at the Planck scale.

REFERENCES

1. P. Langacker,*Proc. PASCOS90*, Eds. P. Nath and S. Reucroft, World Scientific, Singapore(1990); J. Ellis, S. Kelley and D.V. Nanopoulos, *Phys. Lett.* B249:441(1990), B260:131(1991); U. Amaldi, W. DeBoer and H. Furstenau, *Phys. Lett.* B260:447(1991); F. Anselmo, L. Cifarelli, A. Peterman and A. Zichichi, *Nuov. Cim.* 104A:1817(1991); 115A:581 (1992).

2. A.H. Chamseddine, R. Arnowitt and P. Nath, *Phys. Rev. Lett.* 49:970 (1982). For reviews see P. Nath, R. Arnowitt and A.H. Chamseddine, *Applied N = 1 Supergravity*, World Scientific, Singapore(1984); H.P. Nilles, *Phys. Rep.* 100:1(1984); R. Arnowitt and P. Nath, Proc. VII Swieca Summer School, ed. E. Eboli, World Scientific, Singapore(1994).

3. S. Soni and A. Weldon, *Phys. Lett.* B126:215(1983).

4. R. Barbieri, S. Ferrar and C.A. Savoy, *Phys. Lett.* B119:343(1982); L. Hall, J. Lykken and S. Weinberg, *Phys. Rev.* D27:2359 (1983); P. Nath, R. Arnowitt and A.H. Chamseddine, *Nucl. Phys.* B227:121(1983); V. Kaplunovsky and J. Louis, *Phys. Lett.* B306:269(1993).

5. I. Antoniadis, J. Ellis, J. Hagelin and D.V. Nanopoulos, *Phys. Lett.* B208:209(1988).

6. K. Inoue et.al., *Prog. Theor. Phys.* 68:927(1982); L. Ibañez and G.G. Ross, *Phys. Lett.* B110:227(1982); L. Alvarez-Gaumé, J. Polchinski and M.B. Wise, *Nucl. Phys.* B221:495(1983); J. Ellis, J. Hagelin, D.V. Nanopoulos and K. Tamvakis, *Phys. Lett.* B125:2275(1983).

7. E. Ibañez and C. Lopez, *Phys. Lett.* B128:54(1983); *Nucl. Phys.* B233:545(1984); L.E. Ibañez, C. Lopez and C. Muños, *Nucl. Phys.* B256:218(1985).

8. N. Polonsky and A. Pomarol, *Phys. Rev.* D51:6532(1995).

9. R. Arnowitt and P. Nath, hep-ex/9605011; hep-ph/9701325.

10. T. Tsukamoto, K. Fujii, H. Murayama, M. Yamaguchi, and Y. Okada, *Phys. Rev.* D51:3153(1995).

11. J.L. Feng, M.E. Peskin, H. Murayama and X. Tata, *Phys. Rev* D52:1418 (1995).

12. M.M. Nojiri, *Phys. Rev.* D51:6281(1995).

13. M.E. Peskin, talk at YKIS95, Kyoto (1995).

14. J.F. Feng and D.E. Finnell, *Phys. Rev.* D49:2369(1994).

15. H. Haber, *Proc. of Beyond the Standard Model IV*, ed. J. Gunion, T. Hans and J. Ohnemus, World Scientific, Singapore(1995).

16. Y. Kawamura, H. Murayama and M. Yamaguchi, *Phys. Lett.* B324:52(1994).

17. A.E. Faragii, *Phys. Lett.* B278:131(1992); B302:202(1993).

18. M. Dine, R. Leigh and A. Kagan, *Phys. Rev.* D48:4269(1993).

19. A. Font, L.E. Ibañez, D. Lust and F. Quevedo, *Nucl. Phys.* B245:401(1990); M. Cvetič, A. Font, L.E. Ibañez, D. Lust and F. Quevedo, *Nucl. Phys.* B361:194(1991); A. de la Macorra and G.G. Ross, *Nucl. Phys.* B404:321(1993); V. Kaplunovsky and J. Louis, *Phys. Lett.* B306:269(1993); R. Barbieri, J. Louis and M. Moretti, *Phys. Lett.* B312:451(1993); (Err. B316:632(1993)); J.L. Lopez, D.V. Nanopoulos and Z.A. Zichichi, *Phys. Lett.* B319:451(1993); S. Ferrara, C. Kounnas and F. Zwirner, *Nucl. Phys.* B429, 589 (1994) (Err. B433:255(1995)).

20. P. Candelas, M. Lynker and R. Schimmrigk, *Nucl. Phys.* B341:383(1990); J. Fuchs, A. Klemm, C. Scheich and M.G. Schmidt, *Phys. Lett.* B232:317(1989).

21. H. Kim and C. Muños, hep-ph/9608214.

22. A. Klemm and S. Theisen, *Nucl. Phys.* B389:153(1993); A. Font, *Nucl. Phys.* B391:358(1993); S. Hosono, A. Klemm, S. Theisen and S.-T. Yau, *Nucl. Phys.* B433:501 (1995).

ANYONIC BEHAVIOR OF QUANTUM GROUP FERMIONIC AND BOSONIC SYSTEMS

Marcelo R. Ubriaco[1]

Laboratory of Theoretical Physics

Department of Physics

University of Puerto Rico

Río Piedras Campus, P. O. Box 23343

San Juan PR 00931-3343

Introduction

The role of quantum groups and quantum Lie algebras [1] in physics has its origin in the theory of vertex models [2] and the quantum inverse scattering method [3]. From the mathematical point of view, two of the most important developments have been their understanding in terms of the theory of noncommutative Hopf algebras [4] and their relation to non-commutative geometry [5, 6, 7].

In recent years the study of quantum groups and quantum algebras has greatly diversified into several areas of theoretical physics. Based on quantum group ideas, a considerable amount of work was devoted towards a formulation of the so called q-deformed physical systems. These approaches are attempts to develop more general formulations of quantum mechanics [8] and field theory [9, 10]. The main motivation behind this type of projects resides in searching for new roles that quantum groups could play in physics other than the theory of integrable models. A successful and consistent formulation of a theory involving quantum group symmetries will have the potential of having new features no present in the standard $q \rightarrow 1$ case. Besides, it will provide a more general, or alternative, framework to explain physical phenomena.

---

[1]ubriaco@ltp.upr.clu.edu

In this article we show the role that quantum group symmetries, in particular $SU_q(2)$, play in a thermodynamic system at high temperatures. We first display the quantum group covariant algebras, which will be used to build quantum group invariant hamiltonians, and then we will discuss the behavior of the corresponding quantum group gases at high temperatures, and show how the parameter q interpolates between a wide range of attractive and repulsive systems.

Quantum Group Covariant Algebras

As it is well known, boson and fermions operators satisfy

$$\phi_i\phi_j^\dagger - \phi_j^\dagger\phi_i = \delta_{ij}$$
$$\psi_i\psi_j^\dagger + \psi_j^\dagger\psi_i = \delta_{ij}, \tag{1}$$

which, for $i, j = 1, ...N$, are covariant under $SU(N)$ transformations. For the case of unitary quantum group matrices T the coefficients do not commute but satisfy for $N = 2$ the following algebraic relations

$$T = \begin{pmatrix} a & b \\ c & d \end{pmatrix} \tag{2}$$

$$ab = q^{-1}ba \quad , \quad ac = q^{-1}ca$$
$$bc = cb \quad , \quad dc = qcd$$
$$db = qbd \quad , \quad da - ad = (q - q^{-1})bc$$
$$det_q T \equiv ad - q^{-1}bc = 1, \tag{3}$$

with the unitary condition [11] $\bar{a} = d, \bar{b} = q^{-1}c$ and $q \in \mathbf{R}$. Hereafter, we take $0 \le q < \infty$.

A natural question to address is which are the quantum group analogues of Equation (1), which will tell us for example how to build quantum group invariant hamiltonians. The operator algebras covariant under the action of $SU_q(N)$ matrices were given in [12]

$$\Omega_j\bar{\Omega}_i = \delta_{ij} \pm q^{\pm 1}R_{kijl}\bar{\Omega}_l\Omega_k \tag{4}$$

$$\Omega_l\Omega_k = \pm q^{\mp 1}R_{jikl}\Omega_j\Omega_i, \tag{5}$$

where $\Omega = \Phi, \Psi$ and the upper (lower) sign applies to quantum group bosons Φ_i (quantum group fermions Ψ_i) operators. The $N^2 \times N^2$ matrix R_{jikl} is explicitly written as [7]

$$R_{jikl} = \delta_{jk}\delta_{il}(1 + (q - 1)\delta_{ij}) + (q - q^{-1})\delta_{ik}\delta_{jl}\theta(j - i), \tag{6}$$

where $\theta(j - i) = 1$ for $j > i$ and zero otherwise. Denoting the new fields as $\Omega_i' = \sum_{i=1}^{N} T_{ij}\Omega_j$, the $SU_q(N)$ transformation matrix T and the R-matrix satisfy the well known algebraic relations [13]

$$RT_1T_2 = T_2T_1R, \tag{7}$$

and

$$R_{12}R_{13}R_{23} = R_{23}R_{13}R_{12}, \tag{8}$$

with the standard embedding $T_1 = T \otimes 1$, $T_2 = 1 \otimes T \in V \otimes V$ and $(R_{23})_{ijk,i'j'k'} = \delta_{ii'}R_{jk,j'k'} \in V \otimes V \otimes V$.

In particular, for $N = 2$, Equations (4) and (5) are simply written

a) $SU_q(2) - fermions$

$$\{\Psi_2, \overline{\Psi}_2\} = 1 \tag{9}$$
$$\{\Psi_1, \overline{\Psi}_1\} = 1 - (1 - q^{-2})\overline{\Psi}_2\Psi_2 \tag{10}$$

$$\Psi_1\Psi_2 = -q\Psi_2\Psi_1 \tag{11}$$
$$\overline{\Psi}_1\Psi_2 = -q\Psi_2\overline{\Psi}_1 \tag{12}$$
$$\{\Psi_1, \Psi_1\} = 0 = \{\Psi_2, \Psi_2\}, \tag{13}$$

b) $SU_q(2) - bosons$

$$\Phi_2\overline{\Phi}_2 - q^2\overline{\Phi}_2\Phi_2 = 1 \tag{14}$$
$$\Phi_1\overline{\Phi}_1 - q^2\overline{\Phi}_1\Phi_1 = 1 + (q^2 - 1)\overline{\Phi}_2\Phi_2 \tag{15}$$
$$\Phi_2\Phi_1 = q\Phi_1\Phi_2 \tag{16}$$
$$\Phi_2\overline{\Phi}_1 = q\overline{\Phi}_1\Phi_2, \tag{17}$$

which for $q = 1$ become the fermion and boson algebras respectively. These operator relations are very different than those satisfied by the so called q-fermions [14] and q-bosons [15, 16], which are written respectively as

c) q-fermions

$$bb^\dagger + qb^\dagger b = q^N \tag{18}$$
$$b^\dagger b = [N] \tag{19}$$
$$bb^\dagger = [1 - N] \tag{20}$$
$$b^2 = 0 = b^{\dagger 2}, \tag{21}$$

where the bracket $[x] = \frac{q^x - q^{-x}}{q - q^{-1}}$.

d) q-bosons

$$a_i a_i^\dagger - q^{-1}a_i^\dagger a_i = q^N, \quad [a_i, a_j^\dagger] = 0 = [a_i, a_j], . \tag{22}$$

It is simple to check that Equations (18)-(22) are not quantum group covariant, and therefore a quantum group action on the operators b_i and a_j cannot be defined. Hereafter, we discuss the thermodynamic properties of the systems described by the simplest quantum group invariant hamiltonians.

Quantum Group Fermion and Boson Models

Quantum Group Fermion Gas

From Equation (13) we see that for quantum group fermions the occupation numbers are restricted to $m = 0$ or 1, and therefore $SU_q(N)$-fermions satisfy the Pauli exclusion principle. For a given κ, a normalized state is simply written as

$$\overline{\Psi}_2^n \overline{\Psi}_1^m |0\rangle \quad n, m = 0, 1, \tag{23}$$

and the operator $\mathcal{M}_i \equiv \overline{\Psi}_i \Psi_i$ satisfy

$$[\mathcal{M}_2, \Psi_1] = 0 = \mathcal{M}_1 \Psi_2 - q^2 \Psi_2 \mathcal{M}_1. \tag{24}$$

A representation of the Ψ operators in terms of ordinary fermions ψ_j is simply given by the following relations

$$\Psi_m = \psi_m \prod_{l=m+1}^{N} \left(1 + (q^{-1} - 1)M_l\right), \tag{25}$$

$$\overline{\Psi}_m = \psi_m^\dagger \prod_{l=m+1}^{N} \left(1 + (q^{-1} - 1)M_l\right), \tag{26}$$

where $M_l = \psi_l^\dagger \psi_l$.

The simplest Hamiltonian one can write in terms of the operators Ψ_i is simply the one that becomes the free fermion Hamiltonian for $q = 1$. It is given by [17]

$$\mathcal{H}_F = \sum_\kappa \varepsilon_\kappa (\mathcal{M}_{1,\kappa} + \mathcal{M}_{2,\kappa}), \tag{27}$$

where $\mathcal{M}_{i,\kappa} = \overline{\Psi}_{i,\kappa} \Psi_{i,\kappa}$ and $\{\overline{\Psi}_{i,\kappa}, \Psi_{j,\kappa'}\} = 0$ for $\kappa \neq \kappa'$. With use of the fermion representation in Equations (25) and (26), the original Hamiltonian becomes the interacting fermion Hamiltonian

$$\mathcal{H}_F = \sum_\kappa \varepsilon_\kappa \left(M_{1,\kappa} + M_{2,\kappa} + (q^{-2} - 1)M_{1,\kappa}M_{2,\kappa}\right). \tag{28}$$

We see that the parameter $q \neq 1$ mixes the two degrees of freedom in a nontrivial way through a quartic interaction term. The grand partition function for this model is simply written as

$$\mathcal{Z}_F = \prod_\kappa \sum_{n=0}^{1} \sum_{m=0}^{1} e^{-\beta \varepsilon_\kappa (n+m-(1-q^{-2})mn)} e^{\beta\mu(n+m)} \tag{29}$$

$$= \prod_\kappa \left(1 + 2e^{-\beta(\epsilon_\kappa - \mu)} + e^{-\beta\left(\epsilon_\kappa(q^{-2}+1)-2\mu\right)}\right), \tag{30}$$

which for $q = 1$ becomes the square of a single-fermion-type grand partition function. For a high temperature (or low density) gas, we expand the grand partition function $\mathcal{Z}_F$ in terms of the fugacity $z \ll 1$

$$\ln \mathcal{Z}_F = 4V(2m\pi/h^2\beta)^{3/2} \left[\frac{z}{2} - \alpha(q)\frac{z^2}{2} + \gamma(q)\frac{z^3}{3!} + ...\right], \tag{31}$$

where the functions $\alpha(q)$ and $\gamma(q)$ are

$$\alpha(q) = \frac{1}{2^{3/2}} - \frac{1}{2(q^{-2}+1)^{3/2}}$$
$$\gamma(q) = \frac{4}{3^{3/2}} - \frac{3}{(q^{-2}+2)^{3/2}}.$$

Calculating the average number of particles $\langle M \rangle = \frac{1}{\beta}\left(\frac{\partial \ln \mathcal{Z}_F}{\partial \mu}\right)_{T,V}$ and reverting the equation to write the fugacity in terms of $\langle M \rangle$ gives for Equation (31)

$$\ln \mathcal{Z}_F = \langle M \rangle \left[1 + \frac{\alpha(q)\langle M \rangle}{2V}\lambda_T^3 - \frac{\langle M \rangle^2}{16V^2}\lambda_T^6\Lambda + ...\right], \tag{32}$$

where $\Lambda = \frac{8\gamma(q)}{3} + 16\alpha^2(q)$ and $\lambda_T = (h^2\beta/2\pi m)^{1/2}$.

From this equation we can obtain the internal energy $U = -\frac{\partial \ln \mathcal{Z}_F}{\partial \beta} + \mu\langle M \rangle$,

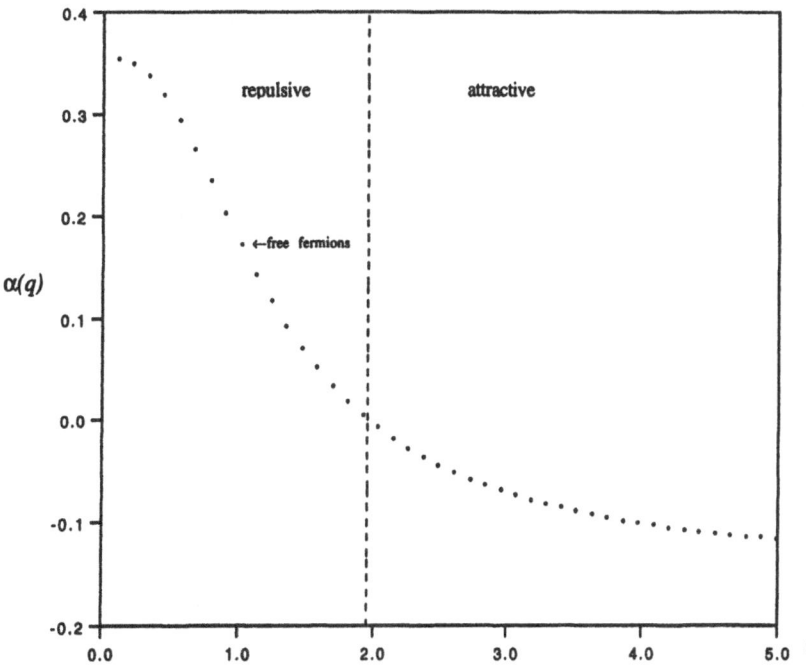

FIG. 1.The coefficient $\alpha(q)$ for the interval $0 \leq q \leq 5$. The line at $q = 1.96$ divides the two regions: $\alpha(q) > 0$ and $\alpha(q) < 0$ which correspond to fermionic and boson-like behavior respectively.

the heat capacity $C_v = \left(\frac{\partial U}{\partial T}\right)_V$ and the entropy $S = \frac{U - \mu \langle M \rangle}{T} + k \ln \mathcal{Z}_F$ as functions of $\langle M \rangle$. The corresponding equations are

$$U = \frac{3\langle N \rangle}{2\beta} \left[1 + \frac{\langle M \rangle}{2V} \lambda_T^3 \alpha(q) - \frac{\langle M \rangle^2}{16V^2} \lambda_T^6 \Lambda + ... \right] \qquad (33)$$

$$C_v = \frac{3\langle M \rangle k}{2} \left[1 - \frac{\langle M \rangle}{4V} \lambda_T^3 \alpha(q) + \frac{\langle M \rangle^2}{8V^2} \lambda_T^6 \Lambda + ... \right], \qquad (34)$$

$$S = \langle M \rangle k \left[\frac{5}{2} - \ln \left(\frac{\langle M \rangle}{2V} \lambda_T^3 \right) + \frac{\langle M \rangle}{4V} \lambda_T^3 \alpha(q) + ... \right]. \qquad (35)$$

The equation of state is given by the equation

$$pV = kT\langle M \rangle \left[1 + \frac{\langle M \rangle}{2V} \lambda_T^3 \alpha(q) + ... \right]. \qquad (36)$$

Clearly, all these functions become, for $q = 1$, the thermodynamic functions for an ideal fermion gas with two species. The sign of the second virial coefficient depends of the value of q, implying then that the parameter q interpolates between repulsive and attractive systems.

Figure 1 shows a graph of the coefficient $\alpha(q)$. The function $\alpha(q)$ takes values in the interval $2^{-5/2} \leq \alpha \leq 2^{-3/2}$ for $0 \leq q \leq 1$, vanishes at $q = 1.96$ and it gets its lowest value $\alpha(q) = -2^{-5/2}(\sqrt{2} - 1)$ in the limit $q \to \infty$. It is important to remark that the second virial coefficient for the ideal boson gas case $B_{bosons} = -2^{-7/2}\beta^{3/2} < B(q \to \infty, T) = -2^{-5/2}(\sqrt{2} - 1)\beta^{3/2}$, and therefore free bosons are not described in this model.

A natural question to address is whether a similar interpolation occurs at $D = 2$. Repeating the previous procedure leads to the equation of state

$$pA = kT\langle M \rangle \left(1 + \frac{1}{4(1 + q^2)} \frac{\langle M \rangle}{A} \lambda_T^2 + ... \right), \qquad (37)$$

wherein the second virial coefficient is positive for all values of q, showing that this model , at $D = 2$, describes only interacting fermionic systems.

Quantum Group Boson Gas

A representation of the quantum group bosons in terms of boson operators ϕ_i and $\phi_j^\dagger$, according to Equations (14)-(17), is simply given by

$$\Phi_2 = (\phi_2^\dagger)^{-1}\{N_2\} \qquad (38)$$
$$\overline{\Phi}_2 = \phi_2^\dagger \qquad (39)$$
$$\Phi_1 = (\phi_1^\dagger)^{-1}\{N_1\}q^{N_2} \qquad (40)$$
$$\overline{\Phi}_1 = \phi_1^\dagger q^{N_2}, \qquad (41)$$

where the bracket $\{x\} = \frac{1 - q^{2x}}{1 - q^2}$ and the boson number operator $N_i = \phi_i^\dagger \phi_i$.

Therefore, the simplest quantum group invariant Hamiltonian [18] $\mathcal{H}_B$

$$\mathcal{H}_B = \sum_\kappa \varepsilon_\kappa (\mathcal{N}_{1,\kappa} + \mathcal{N}_{2,\kappa}), \qquad (42)$$

with $[\overline{\Phi}_{i,\kappa}, \Phi_{j,\kappa'}] = 0$ for $\kappa \neq \kappa'$, becomes the interacting bosonic Hamiltonian

$$\mathcal{H}_B = \sum_\kappa \varepsilon_\kappa \{\phi_{1,\kappa}^\dagger \phi_{1,\kappa} + \phi_{2,\kappa}^\dagger \phi_{2,\kappa}\}, \qquad (43)$$

with the bracket $\{x\}$ as defined below Equation (41). Now, it is simple to write the grand partition function $\mathcal{Z}_B$ for this model. Introducing the chemical potential μ in the usual way gives

$$\mathcal{Z}_B = \prod_\kappa \sum_{n=0}^\infty \sum_{m=0}^\infty e^{-\beta \varepsilon_\kappa \{n+m\}} e^{\beta \mu (n+m)}, \, , \qquad (44)$$

such that after rearrangement of equal power terms it simplifies to the expression

$$\mathcal{Z}_B = \prod_\kappa \sum_{m=0}^\infty (m+1) e^{-\beta \varepsilon_\kappa \{m\}} z^m. \qquad (45)$$

In $D = 3$ the first few terms in powers of z read

$$\begin{aligned}
\ln \mathcal{Z}_B &= \frac{4\pi V}{h^3} \int_0^\infty dp\, p^2 \big(2e^{-\beta \varepsilon_\kappa} z + (6e^{-\beta \varepsilon_\kappa \{2\}} - 4e^{-\beta \varepsilon_\kappa 2}) \frac{z^2}{2} \\
&\quad + (24 e^{-\beta \varepsilon_\kappa \{3\}} - 36 e^{-\beta \varepsilon_\kappa \{2\}} e^{-\beta \varepsilon_\kappa} + 16 e^{-\beta \varepsilon_\kappa 3}) \frac{z^3}{3!} + ... \big), \qquad (46)
\end{aligned}$$

such that performing the elementary integrations gives

$$\ln \mathcal{Z}_B = \frac{4\pi V}{h^3} \left(\frac{\sqrt{\pi}}{2} (\frac{2m}{\beta})^{3/2} z + \sqrt{\pi} (\frac{2m}{\beta})^{3/2} \delta(q) z^2 + ... \right), \qquad (47)$$

where $\delta(q) = \frac{1}{4} \left(\frac{3}{(1+q^2)^{3/2}} - \frac{1}{\sqrt{2}} \right)$.

Calculating the average number of particles $\langle N \rangle = \frac{1}{\beta} \left(\frac{\partial \ln \mathcal{Z}_B}{\partial \mu} \right)_{T,V}$ and reverting the equation we find for the fugacity

$$z \approx \frac{1}{2} \left(\frac{h^2}{2m\pi kT} \right)^{3/2} \frac{\langle N \rangle}{V} - \delta(q) \left(\frac{h^2}{2m\pi kT} \right)^3 \left(\frac{\langle N \rangle}{V} \right)^2. \qquad (48)$$

The internal energy, heat capacity and entropy functions in terms of the average number of particles $\langle N \rangle$ and q read

$$U = \frac{3\langle N \rangle}{2\beta} \left[1 - \frac{\langle N \rangle}{V} \lambda_T^3 \delta(q) + \frac{\langle N \rangle^2}{V^2} \lambda_T^6 \left(4\delta^2(q) - \frac{\Gamma(q)}{12} \right) + ... \right], \qquad (49)$$

$$C_v = \frac{3k\langle N \rangle}{2} \left[1 + \frac{\langle N \rangle}{2V} \lambda_T^3 \delta(q) - 2\frac{\langle N \rangle^2}{V^2} \lambda_T^6 \left(4\delta^2(q) - \frac{\Gamma(q)}{12} \right) + ... \right], \qquad (50)$$

$$S = k\langle N \rangle \left[\frac{5}{2} - \ln \left(\frac{\langle N \rangle}{2V} \lambda_T^3 \right) - \frac{\langle N \rangle}{2V} \lambda_T^3 \delta(q) + ... \right], \qquad (51)$$

where the function $\Gamma(q) = \frac{12}{3^{3/2}} - \frac{18}{(2+q^2)^{3/2}} + \frac{8}{3^{3/2}}$. The equation of state for this model is more interesting than for the $SU_q(2)$ fermion gas. For $D = 3$,

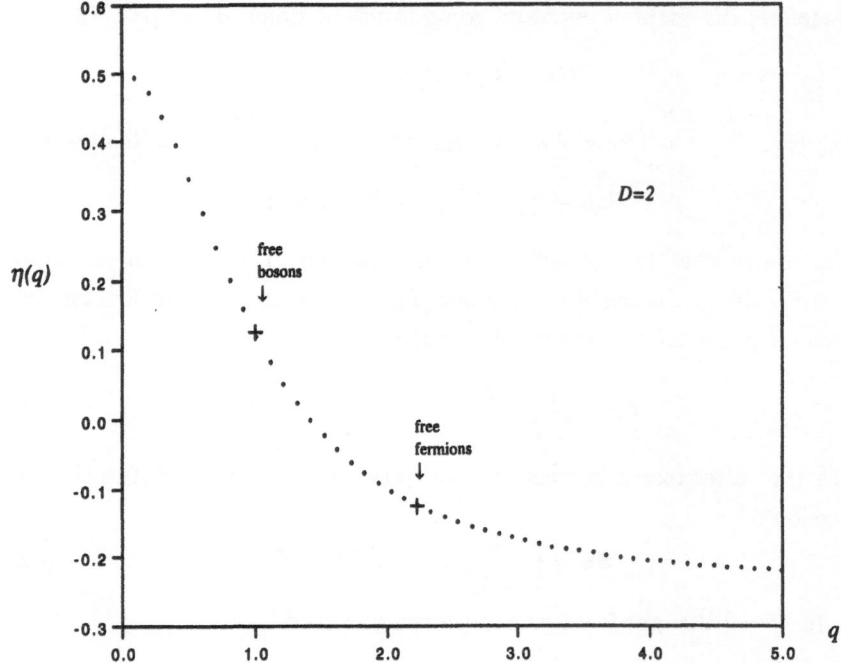

FIG. 2 The coefficient $\eta(q)$ for the interval $0 \leq q \leq 5$. At the values $q = 1$ and $q = 5^{1/2}$ the system behaves as a free boson and fermion gas respectively

the equation of state is given by

$$pV = kT\langle N\rangle \left(1 - \frac{\langle N\rangle}{V}\lambda_T^3 \delta(q) + ...\right). \tag{52}$$

As expected, at $q = 1$ the coefficient $\delta(1) = 2^{-7/2}$, which is the numerical factor in the second virial coefficient for a free boson gas with two species. The free fermion $\delta(q) = -2^{7/2}$ and ideal gas $\delta(q) = 0$ cases are reached at $q \approx 1.78$ and $q \approx 1.27$ respectively.

A similar calculation for $D = 2$ leads to the equation of state

$$pA = kT\langle N\rangle \left(1 - \frac{\langle N\rangle}{A}\lambda_T^2 \eta(q) + ...\right), \tag{53}$$

with $\eta(q) = \frac{(2-q^2)}{4(1+q^2)}$. Figure 2 shows a graph of the coefficient $\eta(q)$ as a function of the parameter q for $D = 2$. The coefficient $\eta(q)$ in Equation (53) takes values in the interval $[-\frac{1}{4}, \frac{1}{2}]$. At $D = 2$ this model behaves as a fermion gas at $q = \sqrt{5}$.

Since the $SU_q(2)$ boson gas at $D = 2$ also interpolates completely between bosons and fermions, we can find a relation between the parameter q and the statistical parameter α for an anyon gas [19] of two species. This relation is given by

$$\alpha = 1 - \sqrt{\frac{5 - q^2}{2(1 + q^2)}}, \qquad (54)$$

where $0 \leq \alpha \leq 1$. The parameter q interpolates within a larger range of attractive and repulsive systems than the α parameter does.

Discussion

In this article we have discussed the high temperature behavior of quantum group gases. We considered the two simplest quantum group invariant Hamiltonians , which are those that become for $q = 1$ the free fermion or boson gases with two species. A representation of the quantum group fermions in terms of ordinary fermions leads to a fermion system with a quartic interaction whose coupling constant vanishes as $q \to 1$. At high temperatures we analyzed the equation of state at $D = 2$ and $D = 3$ spatial dimensions. At $D = 2$ the second virial coefficient is always positive for all values of q, therefore in two dimensional space this model describes only interacting fermion systems. At $D = 3$ the sign of the second virial coefficient depends of the value of q, showing then that the parameter q interpolates between repulsive and attractive behavior. The ideal gas case corresponds to $q = 1.96$ and the system becomes repulsive for $q < 1.96$. For $q > 1.96$ the system becomes attractive, but as $q \to \infty$ the free boson limit is not reached, and therefore this model does not interpolate completely between the free fermion and free boson cases.

For $SU_q(2)$ bosons the results are more interesting. A representation of the quantum group boson operators in terms of ordinary bosons leads to a hamiltonian in terms of ordinary boson interactions involving powers of the number operators and $\ln q$. For $D = 2$ and $D = 3$ the q parameter interpolates completely between the free boson and free fermion cases. For $D = 2$, a comparison with the anyon statistical parameter shows that the parameter q interpolates within a larger range of systems.

Thus, at high temperatures the interactions that result by imposing $SU_q(2)$ symmetry in the simplest hamiltonian are such that these models, and in particular the quantum group boson model, offer an alternative approach in describing systems obeying fractional statistics in two and three spatial dimensions.

References

[1] See for example M. Jimbo, ed., Advanced Series in Mathematical Physics, Vol. 10 *Yang-Baxter equation in integrable systems*, (World Scientific, Singapore, 1990).

[2] H. Saleur and J.-B. Zuber, Integrable lattice models and quantum groups, in: Proceedings of the Trieste Spring School *String Theory and Quantum Gravity*, eds. M. Green, R. Iengo, S. Randjbar-Daemi, E. Sezgin and H. Verlinde, (World Scientific, Singapore, 1992) and references therein.

[3] L. Fadeev, Integrable models in (1+1)- dimensional quantum field theory in *Recent Advances in Field Theory and Statistical Mechanics*, Les Houches 1982, eds. J.-B. Zuber and R. Stora, (North Holland 1984).

[4] V. Drinfeld, Hopf algebras and the quantum Yang-Baxter equation, *Sov. Math. Dokl.* 32:254 (1985).

[5] S. L. Woronowicz, Twisted $SU(2)$ group. An example of a non-commutative differential calculus, *Publ. RIMS* 23:117 (1987).

[6] Yu. I. Manin, Multiparametric quantum deformation of the general linear supergroup, *Comm. Math. Phys.* 123:163 (1989).

[7] J. Wess and B. Zumino,Covariant differential calculus on the quantum hyperplane ,*Nucl. Phys.* B18(Proc. Suppl.):302 (1990).

[8] M. R. Ubriaco, Quantum deformations of quantum mechanics, *Mod. Phys. Lett.* A8:89 (1993); and references therein.

[9] M. R. Ubriaco, Complex q analysis and scalar field theory on a q lattice, *Mod. Phys. Lett.* A9:1121 (1994); and references therein.

[10] A. Sudbery, $SU_q(n)$ gauge theory, *Phys. Lett.* B375:75 (1996).

[11] S. Vokos, B. Zumino and J. Wess, Properties of quantum 2×2 matrices, in: *Symmetry in Nature*, Scuola Normale Superiore Publ., Pisa (1989).

[12] M. R. Ubriaco, Quantum group Schrödinger field theory, *Mod. Phys. Lett.* A8:2213 (1993); A10:2223(E) (1995).

[13] L. A. Takhatajan, Quantum groups and integrable models, *Adv. Stud. Pure Math.*19:1 (1989).

[14] Y.J. Ng, Comment on the q-analogues of the harmonic oscillator, *J. Phys.* A:23, 1203 (1990).

[15] A. J. Macfarlane, On q-analogues of the quantum harmonic oscillator and the quantum group $SU_q(2)$, *J. Phys.* A22:4581 (1989).

[16] L. C. Biedenharn, The quantum group $SU_q(2)$ and a q-analog of the boson operators, *J. Phys.* A22:873 (1989).

[17] M. R. Ubriaco, High and low temperature behavior of a quantum group fermion gas, *Mod. Phys. Lett.* A11:2325 (1996).

[18] M. R. Ubriaco, Anyonic behavior of quantum group gases, *Phys. Rev* E55:291 (1997).

[19] D. Arovas, Topics in fractional statistics, in: *Geometric Phases in Physics*, A. Shapere and F. Wilczek,ed., (World Scientific, Singapore, 1989).

STRINGY *P*-BRANES: SUGGESTIONS OF DIMENSIONAL DUALITY

Gerald B. Cleaver

Department of Physics and Astronomy
The University Of Pennsylvania
Philadelphia, Pennsylvania 19104-6396

ABSTRACT

I discuss how a "dimensional duality" would exist between a pair of string models with $d = (p+1)$ and $d' = (D-p-3)$ non-compact dimensions, if these models could generate, respectively, the zero-modes of dual p- and $p' = (D-p-4)$-branes living in D spacetime dimensions.

String Theory Circa 1995-1996

The epoch of 1995 and 1996 has been labelled the "Second String Revolution" [1]. Driving a change of paradigm in the string field was the gradual understanding (wrought about by several groups' converging lines of research) that the five consistent ten-dimensional string "theories" (Type-I, Type-IIA/B, and $E_8 \times E_8$/SO(32) heterotic), formerly viewed as distinct, were in fact different corners in the same moduli space of a far more encompassing theory. In like manner, the string "theories" were further connected to 11-dimensional membrane theory, which yields 11-dimensional supergravity as its low energy limit. One candidate for the unified theory has been entitled (eleven-dimensional) "M"(embrane, atrix, onster, ystery, ...)-theory [2, 3], while another proposal is twelve-dimensional "F"-theory [4, 5, 6], involving two time-like directions. The various regions of the moduli space corresponding to ten-dimensional string "theories" and eleven-dimensional membrane "theory" overlap via strong/weak dualities. When there is $N \geq 2$ (with $N = 1$ sometimes sufficient) spacetime (ST) supersymmetry (SUSY), a strongly coupled model of one theory, e.g. heterotic, is phenomenologically equivalent to a weakly coupled model of an alternative "theory", e.g. Type-I.

A key element of this revolution was the understanding that strings (one-branes) are not the only higher-dimensional objects that arise in "string theory," but that additional $(p > 1)$-branes do as well. *P*-branes, as a whole, are now viewed much more democratically [8, 9], with the string far less unique then previously envisioned.

Under strong/weak duality transformations between string models, two physical properties have been conserved: (1) the number of spacetime supersymmetries, and (2) the number of non-compact dimensions of spacetime. While it is thought that spacetime supersymmetry (ST-SUSY) is indeed a constant of dual string models, I would suggest in contrast to this, that the dimension of non-compact spacetime may not always be conserved. Certain combinations of dualities could perhaps lead from a string model embedded in d non-compact spacetime dimensions to a string model embedded in a different number of d' non-compact spacetime dimensions. I suggest that a such a *dimensional duality* would be deeply related to the duality [10, 11] between (super) p-branes [3, 12] and (super) $(D - p - 4)$-branes embedded in a D-dimensional spacetime. Implications of this dimensional duality, were it found to exist, could be profound. It could perhaps lead to some understanding of the non-perturbative dynamics responsible for producing exactly four non-compact dimensions in this physical universe.

In this talk will focus on the concepts involved in this possible dimensional duality. In section two I give a brief review of generic (super) p-branes and their zero modes. Next, in section three I examine p-brane duality. In section four I discuss the findings of Kutasov et al. [4], who have demonstrated that both string and two-brane dynamics may be described by $N = (2,1)$ heterotic string models. I then suggest that generalization of the ideas of Kutasov et al., *i.e.*, using strings to furnish descriptions of $(p \geq 2)$-branes, might produce examples of this dimensional duality. Last, in section five I discuss my own progress in attempting to find an example of dimensional duality. I review my current efforts to connect a string model in $d = 3$ non-compact dimensions to another in $d' = 6$ non-compact dimensions, through the associated duality between $p = 2$ and $p' = 5$-branes in $D = 11$ spacetime.

P-Branes and P-Brane Zero-Modes

P-branes are generalizations of strings. Whereas a string is a one-dimensional object that sweeps out a $(d = 1 + 1)$-dimensional worldsheet as it evolves in time, generic p-branes are p-dimensional objects that sweep out $(d = p + 1)$-dimensional worldvolumes. While p-branes were first envisioned as fundamental objects by Dirac (for $p = 2$) in 1962 [13], supersymmetric extensions of non-string p-branes were first proposed in 1986 by Hughes, Liu, and Polchinski [14]. The only classical constraint on p-branes in D-dimensional spacetime, is $p < D$. However, supersymmetrizing places strong constraints on allowed values of both p and D. Classification of consistent super p-branes has been performed, resulting in lists of classically allowed super p-branes aptly referred to as *brane-scans* [15, 3, 12].

The trajectory of a p-brane, as it moves through a D-dimensional spacetime can be labelled by scalar functions $X^M(\xi)$, where $\xi^{i=0 \text{ to } p}$ are the worldvolume coordinates. In *static gauge*, the spacetime coordinates are divided into two groups, with d and $D - d$ components, respectively:

$$X^M(\xi) = \left(X^{\mu=0 \text{ to } p}(\xi), X^{m=p+1 \text{ to } D}(\xi) \right) . \tag{1}$$

The worldvolume coordinates of the p-brane are then identified with the first (*i.e.*, d-dimensional) set of spacetime coordinates,

$$X^\mu(\xi) = \xi^\mu . \tag{2}$$

This leaves the $D - d$ coordinates X^m as the on-shell physical scalar degrees of freedom (DOF) of the p-brane. The X^m are the directions transverse to the p-brane and correspond to Goldstone zero-modes produced by the brane breaking the global translational symmetries along these directions. In addition, various p-branes may have other bosonic worldvolume DOF. Denoting the number of additional bosonic zero-modes by N_B^{extra}, the total number of bosonic DOF, N_B, is therefore

$$N_B = (D - d) + N_B^{extra} . \tag{3}$$

Supersymmetrizing a p-brane augments the $X^M(\xi)$ coordinates with anticommuting fermionic coordinates $\theta(\xi)$. Whether the spinors are Dirac, Weyl, Majorana, or Majorana-Weyl depends on the spacetime dimension D. Kappa symmetry removes half of the spinor DOF on the worldvolume, while going on-shell removes half again. Therefore, if in D-dimensional spacetime the number of a minimal spinor's real components is M and the number of supersymmetries is N (and m and n are the corresponding quantities on a p-brane's d-dimensional worldvolume), the total number of on-shell fermionic DOF is

$$N_F = \frac{1}{2} m n = \frac{1}{4} M N . \tag{4}$$

The worldvolume supersymmetry requirement $N_B = N_F$ for a super p-brane, therefore, translates into

$$(D - d) + N_B^{extra} = \frac{1}{2} m n = \frac{1}{4} M N. \tag{5}$$

Various brane-scans have determined the solutions to eq. (5), under the constraint of using only states with spins up to two as sources of additional bosonic zero-modes. ([16] provides an exhaustive classification of all unitary massless supermultiplets with spins in this range.) Table 2 shows the latest brane-scan results [3].

The only allowed exception to eq. (5) is the heterotic string. Since left- and right-moving modes of a string ($d = 2$) are independent only one of the two sets of modes need satisfy $N_B = N_F$. (Also note that $N_B^{extra} = 0$ for all strings.) Thus, in the case of the heterotic string

$$(D - 2) = n = \frac{1}{2} M N . \tag{6}$$

Consider the possible super p-branes in $D = 11$ spacetime. In this dimension there is but a single (M, N) combination (*i.e.* $M = 32$, $N = 1$), so any such a superbrane must have exactly eight bosonic (fermionic) zero-modes. Only two (possibly three) related super $p-$brane solutions have been found: a super two-brane (with $m = 2$, $n = 8$), a super five-brane (with $m = 8$, $n = 2$), and perhaps a super nine-brane (see [3]). The super two-brane provides the eight bosonic modes directly through its $8 = D - d = 11 - 3$ transverse DOF X^M [17]. It has no other classes of bosonic zero-modes (with one caveat from dimensional compactification discussed below).

On the other hand, the super five-brane has only $5 = D - d' = 11 - 6$ transverse bosonic modes: the missing three come from a rank two antisymmetric tensor (AST) $B_{\mu\nu}^-$ on the worldvolume [18, 19, 3]. Together these $5 + 3 = 8$ zero-modes form the physical states of an $N = 2$ AST supermultiplet. Interestingly, the $N = 2$ AST is formed from the sum of an $N = 1$ AST multiplet, containing $B_{\mu\nu}^-$ and a single scalar, and an $N = 1$ hypermultiplet containing four scalars. Therefore, one (combination) of the transverse Goldstone modes can be distinguished from the others by its $N = 1$ association with $B_{\mu\nu}^-$.)

Table 1. Minimal spinor components and supersymmetries (borrowed from [3]).

Dimension (D or d)	Minimal Spinor (M or m)	Supersymmetry (N or n)
11	32	1
10	16	2, 1
9	16	2, 1
8	16	2, 1
7	16	2, 1
6	8	4, 3, 2, 1
5	8	4, 3, 2, 1
4	4	8, ..., 1
3	2	16, ..., 1
2	1	32, ..., 1

Table 2. The brane scan, where S, V and T denote scalar, vector and antisymmetric tensor multiplets solutions (borrowed from [3]).

$D\uparrow$	0	1	2	3	4	5	6	7	8	9	10	11
11	.			S			T			?		
10	.	V	S/V	V	V	V	S/V	V	V	V		
9	.	S				S						
8	.			S								
7	.			S			T					
6	.	V	S/V	V	S/V	V	V					
5	.	S		S								
4	.	V	S/V	S/V	V							
3	.	S/V	S/V	V								
2	.	S										
1	.											
0	.	.	.	.	.	.	.	.	.	.	.	$d\rightarrow$

P-Brane Duality

A given p-brane may exist as an *elementary particle* (a.k.a. *fundamental particle*) or as a *soliton* (a.k.a. *topological defect*) [20]. Elementary generally means the brane "carries a non-vanishing Noether electric charge e_p following from the equations of motion," while as a soliton the brane is a "source free solution of the field equations which carries a topological magnetic charge m_p" [3]. (If a p-brane carries both electric and magnetic charge it is termed a *dyon*.) If a given action S in D-dimensional spacetime admits a certain (super) p-brane as an elementary solution, then the same action will also admit a corresponding (super) $(p' = D - p - 4)$-brane as a solitonic solution. Such p- and p'-branes are denoted as *dual branes*. The two-brane and five-brane[21] are examples of duals in $D = 11$-spacetime, Correspondingly, the one-brane and five-brane are ten-dimensional duals [22], as are the two-brane and four-brane, while the three-brane is self-dual [23].

There is a Dirac quantization rule relating the electric and magnetic charges of the dual branes charges[10, 11],

$$e_p m_{p'} = 2\pi n; \quad \text{where} \quad n = \text{Integer} \tag{7}$$

This quantization can be reexpressed for branes in terms of a relation between tensions T_p and $T_{p'}$ [24],

$$2\kappa^2 T_p T_{p'} = 2\pi n . \tag{8}$$

where κ is the D-dimensional gravitational constant.

There is (or is expected to be) a physically equivalent dual action $\tilde{S}$ that reverses the roles of the p- and p'-branes, accepting the p-brane as a solitonic solution and the p'-brane as the fundamental. Thus, electric and magnetic charges are interchanged [25],

$$\tilde{m}_p = e_p \quad \text{and} \quad \tilde{e}_{p'} = m_{p'} . \tag{9}$$

(Though this argument says the dual action $\tilde{S}$ exists, this does not mean that one action can automatically be derived one from the other. For example, although the Green-Schwarz action for the $D = 11$ super two-brane has been around for some time, only in the last few months has significant progress been made in deriving actions and field equations for the dual super five-brane [26, 27, 28, 29].)

Stringy Descriptions of P-Branes

Kutasov, Martinec, and O'Loughlin [4] have recently revived and extended an old idea of Green's [30] that a string theory embedded in an effective $(1 + 1)$-dimensional target spacetime has a low-energy effective action that is the worldsheet field theory of another string. The key to this approach is to reinterpret the non-compact spacetime of the initial string as the worldsheet of another string. Massless spacetime states of the first string become the worldsheet zero-modes of another string. While Green was able to primarily describe non-critical strings from this redefinition, Kutasov et al. discovered that increasing the worldsheet supersymmetry of the initial string tends to yield descriptions of other strings closer to criticality. Starting with $N = 2$ strings [31], proved to be the secret to describing critical strings. Specifically, strings with $N = (2, 1)$ or $N = (2, 0)$ worldsheet SUSY produced success; $N = (2, 2)$ strings were too constraining. Upon compactifying $N = 2$ strings in critical $(2+2)$ spacetime to live in an effective $(1+1)$ target space, Kutasov et al. were able to describe the whole gamut of strings classes: Type-IIA/B, Type I/I', SO(32)/$E_8 \times E_8$ heterotic, and bosonic.

Amazingly, Kutasov et al. also discovered that strings could describe more than just other strings! For proper choice of internal field boundary conditions, an $N = (2, 1)$

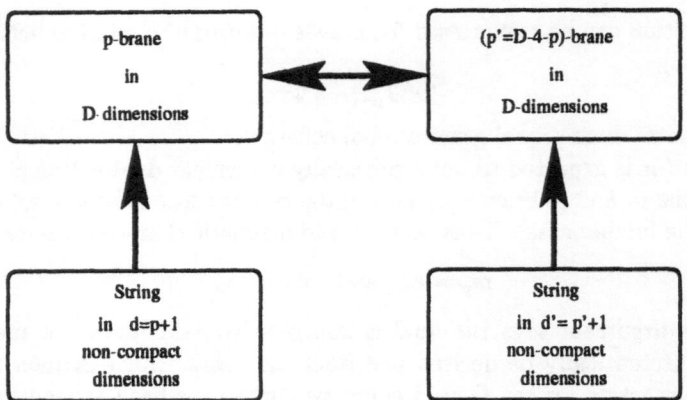

Figure 1. Possible Stringy "Dimensional Duality" from Dual Branes.

string can yield the zero modes of a *supermembrane* in $(10 + 1)$-dimensional space as well. Rather than compactifying from $(2+2)$ down to $(1+1)$ spacetime, compactification to $(2 + 1)$ is performed instead. This allows non-compact spacetime's redefinition as a worldvolume rather than as a worldsheet.

The supermembrane's eight physical bosonic zero modes [15] were formed from a set of seven scalars $V^{a=1 \text{ to } 7}$ and a three dimensional gauge field A^μ [4]. (Supersymmetry's accompanying eight fermionic zero-modes appear as the only other massless states in the spectrum besides the supergravity multiplet.) In $(2 + 1)$ dimensions the gauge field has one propagating degree of freedom and is dual to a scalar via $\epsilon^{\mu\nu\rho} F_{\nu\rho} = \partial^\mu \phi$. This combination $7 + 1$ of bosons, rather than simply being eight scalars, implies compactification of one of the spatial dimensions of $(10 + 1)$ spacetime. That is, dualizing a scalar into a vector can be interpreted as eliminating an effective transverse direction of spacetime through compactification. (On the other hand, a parallel dualization does not simultaneously occur upon compactification for any one of the five-brane scalar modes. The $D = 11$ five-brane discussed above was, in fact, first viewed as the $D = 10$ Type-IIA five-brane.[18])

Kutasov et al. pointed out that a stringy description of membrane dynamics has significant implications. For example, since the $(2 + 1)$ model is a restricted form of $2+2$ self-dual gravity coupled to self-dual matter, the classical theory (given membrane interpretation) must be integrable [4]. Further, the $(2, 10)$ string gives a prescription for the quantum theory of the membrane. Kutasov et al. recognized the possibility of directly quantizing the target space effective field theory and, thus, yielding a promising candidate for a consistent quantum theory of supermembranes in 11 dimensions.

Stringy Dimensional Duality Via Dual P-Branes?

Might it be possible to construct string models that mimic higher dimensional branes than just the membrane? If so, an alternative to an $N = 2$ string setting is probably necessary. Under spacetime $\leftrightarrow$ worldvolume duality, having $D = 4$ as the critical spacetime dimension (with either $(4+0)$ or $(2+2)$ signature) for $N = 2$ strings imposes a tight constraint on the range of worldvolumes approachable. Thus, one probably must start with strings in larger critical dimensions. Of particular interest would be a stringy description of super five-branes, the eleven dimensional dual of the supermembrane. To yield a consistent superstring theory with an effective six-dimensional target space (reinterpretable as the worldvolume of a five-brane) compactification of a ten dimensional theory is most-likely required.

A stringy description of five-brane zero-modes could, of course, yield a candidate for a consistent super five-brane quantum theory in 11 dimensions. Perhaps this might compliment recent progress that has followed more direct lines of attack [26, 27, 28]. But much more might result as well! I would suggest that $D = 11$ supermembrane/super five-brane duality might imply that two distinct string theories yielding membranes and five-branes zero-modes, respectively, should also be viewed as dual models. This would present an example of a string duality that obviously would not conserve the dimension of non-compact spacetime. That is, one string model would have $d = p + 1$ non-compact spacetime dimensions while its dual would have $d' = p' + 1$, where p and $p' = D - 4 - p$ are the dimensions of the corresponding dual branes in D-dimensional spacetime. (See Figure 1.)

I am in the process of searching for evidence that a stringy dimensional duality does, indeed, exist. As a starting point, I am investigating whether a stringy five-brane model dual to the Kutasov et al. stringy membrane model could be formed from Type-II strings. In particular, I am using free fermionic construction, while initially involving only periodic/antiperiodic fermions. Although I have not as yet found a model with states corresponding to the zero-modes of exactly one five-brane, I have found that models with varying numbers of five-brane zero-mode sets are possible [32]. The simplest of these models contains 20 $N = 2$ AST multiplets (RR class). This model has four sector generators (*i.e.* basis vectors): the all-periodic sector, a left-moving (LM) SUSY generator, a right-moving (RM) SUSY generator, and a sector that simultaneously reduces the supersymmetry of each SUSY generator from $N = 2$ to $N = 1$. (Mod trivial permutations of the internal fermions, the set of fermion boundary conditions of these generators is unique.)

Variations on these models appear to be able to contain reduced numbers of AST multiplets: for example twelve, eight, and four. Detailed analysis of these models is underway. Other numbers of copies of $N = 2$ AST are probably possible. I have, however, discovered a generic difficulty in producing exactly one set of five-brane zero modes from Type-II free fermionic strings. This is a result of SO(4) global rotational symmetries among the 12 LM and the 12 RM real internal fermions, respectively, that cannot be totally broken. This symmetry produces multiple copies of a given $N = 2$ AST multiplet.

Each set of LM and RM fermions can be separated into four groups of three, with each group corresponding to a compactified dimension and the associated fermion. Label these sets $(x, y, w)^{i=1\ \text{to}\ 4}$ and $(\bar{x}, \bar{y}, \bar{w})^{\bar{i}=1\ \text{to}\ 4}$ for the LM and RM, respectively. In the simplest model there is an SO(4) global rotational symmetry on the index i (and likewise on $\bar{i}$). Assign the periodic internal fermions for the LM and RM SUSY generators to be the x^i and $\bar{x}^{\bar{i}}$, respectively. Then, although the index symmetry may be broken for the y^i and w^i (and similarly for the $\bar{y}^{\bar{i}}$ and $\bar{w}^{\bar{i}}$), through additional basis vectors, this symmetry cannot be broken for the x^i or the $\bar{x}^{\bar{i}}$ if the $N_{LM} = 1$ and $N_{RM} = 1$ supersymmetries are to remain unbroken. Whether this prevents a single $N = 2$ AST multiplet from every being produced under this construction remains to be seen. If ultimately it does not prove possible, other methods of compactification for Type-II strings will be tried followed by conduction of a similar investigation using Type-I$^{(')}$ strings instead.

Concluding Comments

I have considered how a *dimensional duality* may exist between some string models living in different non-compact spacetime dimensions $d = (p+1)$ and $d' = (D-p-3)$. Dimensional duality would be implied by stringy descriptions of dual p- and $p' = (D-p-4)$-branes living in D-dimensional spacetime. This duality, if found, could perhaps lead to an understanding of why there are exactly four non-compact dimensions in a universe with at least a ten-dimensional spacetime.

ACKNOWLEDGMENTS

G.C. wishes to thank the organizers of ORBIS SCIENTIAE 1997, in particular Behram N. Kursunoglu, for producing such a stimulating and enjoyable conference. G.C. also thanks Burt Ovrut for suggesting an investigation into stringy descriptions of generic p-branes, beyond $p = 1$ or 2.

References

1. J. Schwarz, *The Second String Revolution*, hep-th/9607067.

2. For general reviews of M-Theory see for example:
   J. Schwarz, *Lectures on Superstrings and M Theory Dualities*, hep-th/9607201;
   M. Duff, *M-Theory (The Theory Formerly Known As Strings)*, hep-th/9608117;
   J. Polchinski, *TASI Lectures on D-Branes*, hep-th/9611050;
   P. Townsend, *Four Lectures on M-Theory*, hep-th/9612121;
   K. Stelle, *Lectures on Supergravity P-branes*, hep-th/9701088.

3. M. Duff, *Supermembranes*, CTP-TAMU-nn/96; hep-th/9611203.

4. D. Kutasov and E. Martinec, *New Principles for String/Membrane Unification*,
   Nucl. Phys. **B477** (1996) 652;
   D. Kutasov, E. Martinec, and M. O'Loughlin, *Vacua of M-Theory and $N = 2$
   Strings*, Nucl. Phys. **B477** (1996) 675;
   D. Kutasov, E. Martinec, *M-Branes and N=2 Strings*, hep-th/9612102.

5. C. Vafa, *Evidence for F-Theory*, hep-th/9602022.

6. S. Hewson and M. Perry, *The Twelve Dimensional Super (2+2)-Brane*, hep-th/9612008.

7. P. Horava and E. Witten, *Heterotic and TYpe I String Dynamics from Eleven
   Dimensions*, Nucl. Phys. **B460** (1996) 506.

8. I. Bars, *Is There a Unique Consistent Theory Based on Extended Objects. USC-88/HEP06*.

9. P. Townsend, *P-Brane Democracy*, hep-th/9507048.

10. R. Nepomechie, *Magnetic Monopoles from Antisymmetric Tensor Gauge Fields*,
    Phys. Rev. **D31** (1985) 1921.

11. C. Teitelboim, *Monopoles of Higher Rank*, Phys. Lett. **176B** (1986) 69.

12. P. Howe and E. Sezgin, *Superbranes*, CERN-TH/96-200; kcl-th-96; CTP TAMU-28/96; IC/96/126; hep-th/9607227.

13. P. Dirac, Proc. Roy. Soc. **A268** (1962) 57.

14. J. Hughes, Liu, and J. Polchinski, *Supermembranes*, Phys. Lett. **108B** (1986) 370.

15. M. Duff and J.X. Lu, *Type II P-Branes: The Brain-Scan Revisited*,
    Nucl. Phys. **B390** (1993) 276.

16. J. Strathdee, *Extended Poincaré Supersymmetry*, Int. J. Math. Phys. **2** (1987) 273.

17. M. Duff and K. Stelle, *Multimembrane Solutions of $d = 11$ Supergravity*,
    Phys. Lett. **253B** (1991) 113.

18. C. Callan, J. Harvey, and A. Strominger, *Worldbrane Actions for String Solitons*,
    Nucl. Phys. **B367** (1991) 60.

19. D. Kaplan and J. Michelson, *Zero Modes for the $D = 11$ Membrane and Five-Brane*, hep-th/9510052.

20. R. Rajaraman, *Solitons and Instantons*, North Holland, Amsterdam, 1982.

21. P. Howe and E. Sezgin, $D = 11, p = 5$, CERN-TH/96-16; kcl-th-96; CTP TAMU-55/96; hep-th/9611008.

22. A. Strominger, *Heterotic Solitons*, Nucl. Phys. **B343** (1990) 167.

23. M. Duff and J. Lu, *The Self-Dual Type IIB Superthreebrane*, Phys. Lett. **273B** (1991) 409.

24. M. Duff, and J. Lu, *Remarks on String/Fivebrane Duality*, Nucl. Phys. **B354** (1991) 141.

25. A. Strominger. *Heterotic Solitons*, Nucl. Phys. **B343** (1990) 167.

26. M. Aganagic, J. Park, C. Popescu, and J. Schwarz, *World-Volume Action of the M Theory Five-Brane*, CALT-68-2093 hep-th/9701166.

27. P. Pasti, D. Sorokin, and M. Tonin *Covariant Action for a D=11 Five-Brane with the Chiral Field*, hep-th/9701037.

28. P. Howe and E. Sezgin, and P. West, *Covariant Field Equations of the M Theory Five-Brane* CERN-TH/96-16; kcl-th-96; CTP TAMU-55/96; hep-th/9611008.

29. S. Hewson, *Generalized Supersymmetry and P-Brane Actions*, hep-th/9701011.

30. M. Green, Nucl. Phys. **B293** (1987) 593.

31. H. Ooguri and C. Vafa, Mod. Phys. Lett. **A5** (1990) 1389;
Ibid, Nucl. Phys. **B361** (1991) 469; Nucl. Phys. **B367** (1991) 83.

32. G. Cleaver, work in progress.

SECTION IV

ROUND TRIP BETWEEN COSMOLOGY AND ELEMENTARY PARTICLES

PHYSICS OF MASS

Behram N. Kursunoglu

Global Foundation, Inc., Coral Gables, Florida
(kursungf@netrunner.net)

In the general relativistic theory of gravitation, Schwarzschild singularity yields the mass relation

$$M = \frac{c^2}{2G} \, r \tag{1}$$

where r is the Schwarzschild radius lying well inside the particle of mass M. In the author's version of the generalized theory of gravitation[4] the equation of state

$$q^2 \, r_o^{\,2} = \frac{c^4}{2G} \,, \tag{2}$$

yields the mass relation

$$M = \frac{c^2}{2G} \, r_o \,, \tag{3}$$

where r_o is a fundamental length and is independent of the coordinates. The mass M is defined by multiplying the energy density q^2 by the cubic volume $r_o^{\,3}$. The non-symmetric field variables

$$\hat{g}_{\mu\nu} = g_{\mu\nu} + q^{-1}\Phi_{\mu\nu} \,, \tag{4}$$

of the theory in the limit $r_o = 0$ or $q = \infty$ reduces to the symmetric field $g_{\mu\nu}$ of the general theory of relativity.

In the generalized theory of gravitation a new concept of magnetic charge plays the most fundamental role where the magnetic charges g_n (n = 1, 2, 3,...) are confined according to

$$e_n^{\,2} + g_n^{\,2} = Q_n^{\,2}, \tag{5}$$

where the electric charges e_n have the same sign while the signs of g_n alternate

i.e., $g_n = (-1)^n \, |g_n|$ and further that $|g_{n+1}| < |g_n|$ or $\displaystyle \lim_{n \to \infty} \, g_n \to 0$.

The total charge Q_n is given by

$$Q_n^{\,2} = \frac{c^4 \, r_{on}^{\,2}}{2GN^2} \,, \tag{6}$$

where N^2 refers to the ratio of the gravitational force to the sum of the electric and magnetic charge generated forces and where the fundamental length r_o had first been obtained in 1951 during my graduate student days in Cambridge University while modifying Einstein's and Schrödinger's formulations of the non-symmetric generalization of the general relativistic theory of gravitation to obtain the latter in the limit of $r_o = 0$.

High Energy Physics and Cosmology
Edited by B.N. Kursunoglu *et al.*, Plenum Press, New York, 1997

Dirac's relation for a free monopole is given by

$$eg = \pm \tfrac{1}{2} n\hbar c, \quad n = 0, 1, 2, 3\ldots, \tag{7}$$

where the restriction imposed on the electric charge e by a finite magnetic charge g was interpreted by Dirac as a quantization of the electric charge. For the spectra of the e and g values, the graphic form of the relation (7) is represented in Fig. 1 below.

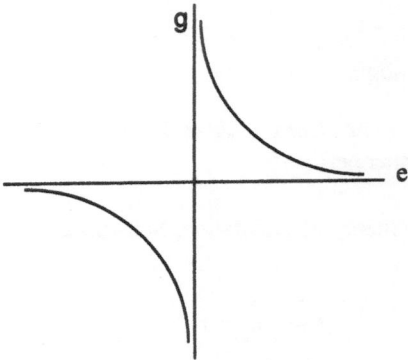

Fig. 1 - Dirac's free monopole behavior relative to electric charge.

The author's generalized theory of gravitation yields as shown in the figure below confined distribution of magnetic charges where there are no free monopoles.

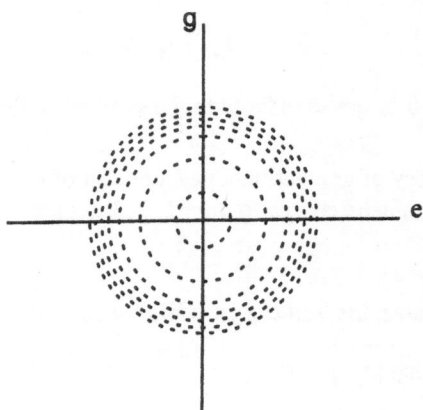

Fig. 2 - In this case, the magnetic charges g_n have alternating signs and they are confined to constitute the structure of a quark or just a particle. The circles correspond to different values of the charge Q_n.

In the Fig. 2, each magnetic charge is represented by a point on a circle placed in the upper first quarter or lower fourth quarter with the left side of the g axis corresponding to an antiparticle.

The fundamental *running constant* r_0 belongs to a non-linear eigen-value of the spherically symmetric field equations[(2)] of the author's version of the generalized theory of gravitation. Like the quantum mechanical description of the hydrogen atoms' energy levels the running length r_0 takes on descrete values. Just for the sake of a preliminary illustration, I would like to discuss, briefly, the mass spectrum based on the gravitational mass and the gravitational size of the elementary particles and all other gravitational masses whose constituents are elementary particles. First let us note that Planck mass is the only mass-scale in physics which we can use to express the most general gravitational mass spectrum,

$$M = \frac{c^2}{2G} \, r_0, \tag{8}$$

of this theory in the form

$$M = \tfrac{1}{2} \mathcal{R} \, m_p, \tag{9}$$

where m_p represents Planck mass

$$m_p = \sqrt{\left(\frac{\hbar c}{G}\right)}, \tag{10}$$

and where $\mathcal{R}$ represents the ratio of the gravitational size r_0 of a particle to the Planck length r_p, viz.,

$$\mathcal{R} = \frac{r_0}{r_p}, \quad r_p = \sqrt{\left(\frac{\hbar G}{c^3}\right)}. \tag{11}$$

Examples of mass values are $\mathcal{R} = 1$, $\mathcal{R} \cong 10^{-20}$, $\mathcal{R} \cong 10^{60}$ yielding the masses $\tfrac{1}{2} m_p$, m (proton), and m (universe) $\cong 8 \times 10^{22}$ solar masses, respectively. The gravitational size of the present theory corresponds to the Schwarzschild radius of general relativity which falls inside the particle. However, here the quantity r_0 is independent of the coordinates and, therefore, is not related to Schwarzschild singularity. The choice of r_0 being of the order of 3 km yields from (8) or (9) correct order of magnitude (2×10^{33}g.) for the solar mass. In fact, $\mathcal{R} \geq 1$ may correspond to the masses of black holes. The corresponding gravitational sizes are: r_0 (Planck) $= r_p$, r_0 (proton) $\cong 10^{-53}$ cm, and r_0 (universe) $\cong 2.58 \times 10^{10}$ light years $\cong 10^{27}$ cm, respectively. Thus the ratio of the mass of the universe to the mass of a proton is of the order of 10^{80} yielding the ballpark value. If we assume that all of the matter content of the expanding universe is converted into electromagnetic waves carrying energy and momentum, then its present age $\tau_0 = (1/c) \, r_0$ (universe) $\cong 25.8 \times 10^9$ years, which is too old compared to the usual age lying between 10 and 20 billion years. I must point out that the fundamental length r_0, when it assumes the size of the universe, the field equations yields approximately flat space-time proving clearly that the universe is flat!

REFERENCES

1. Behram N. Kursunoglu, *Unified Symmetry In the Small and In the Large*, 1995, Volumes 1 and 2, Plenum Press, New York, edited by Behram N. Kursunoglu et al.

2. Behram N. Kursunoglu, *Journal of Physics Essays,* Vol. 1, No. 4, pp. 439-518, 1991, University of Toronto Press.

3. P.A.M. Dirac, *Proceedings of Royal Society,* A133, 60 (1931); idem, *Physical Review*, 13D, 1538 (1976).

4. Behram N. Kursunoglu, *Physical Review*, 88, 1369 (1952).

5. Behram N. Kursunoglu, *Physical Review* D, Volume 12, Number 6, 15 March 1976.

PROGRESS AND PROSPECTS IN THE DIRECT SEARCH FOR SUPERSYMMETRIC DARK MATTER PARTICLES

David B. Cline

Department of Physics and Astronomy, Box 951547
University of California Los Angeles
Los Angeles, CA 90095-1547, USA

ABSTRACT

We briefly review the latest evidence for non-baryonic dark matter. We also review the most recent estimates for SUSY dark-matter particle interaction with matter. We then provide a critique of the current search for the dark matter and describe one very promising method using background discrimination in liquid xenon.

INTRODUCTION

The latest evidence for dark matter in the Universe has been reviewed recently at two University of California Los Angeles (UCLA) symposiums.[1] Remarkably, even in the 1920s some evidence had been found and of course in the 1930s, F. Zwicky provided perhaps the first definitive evidence for dark or non-luminous matter in galaxies.[2]

While no one knows the exact cause of dark matter, there is a reasonable likelihood that new elementary particles play some role in this phenomenon. Of all of the current ideas in this regard, many feel supersymmetry (SUSY) is the most "natural." Our viewpoint is to take the SUSY model seriously and to see what level of detection and discrimination is required to observe such particles. While even the SUSY model is not fully predictive, it would appear to be better than other even more ad hoc models. The project described here grew out of the ICARUS project to construct a massive "electronic bubble chamber" using liquid argon.[3] The first stage of this project, the construction of a 600-ton detector for Hall C at the Gran Sasso, is now approved.

EVIDENCE FOR NON-BARYONIC DARK MATTER

The evidence for non-baryonic dark matter is getting even stronger than it has been. Perhaps the best evidence comes from three different measurements of the gravitational potential in galactic clusters using
1. Dynamic methods (galaxy motion),
2. Hot X-rays (hot gas in the gravitational potential well),
3. Weak gravitational lensing (sensitive to all gravitating matter).

These three methods now give convincing evidence for dominant non-baryonic dark matter in these clusters.

There is also strong evidence for dark matter in dwarf galaxies, which appear to totally dominate the total matter on all scales. Finally, it is now agreed that $\Omega > 0.4$, which is definitely larger than the limits for $\Omega_B < 0.1$ from recent nucleosynthesis studies (see Fig. 1).

METHODS AND RATES TO DETECT SUSY WIMPS

There are many estimates for the cross section of SUSY WIMPs with various targets. We believe this illustrates the difficulty, as well as the promise, for the search for SUSY WIMPs. In this report, we follow the recent work of Nath and Arnowitt[4] (and the references cited therein). Without getting into the details of the assumptions in this calculation, we note that the range of rates goes from a few events/kg·d to 10^{-5} events/kg·d (see Figs. 2 and 3). Although the results are for Ge and Pb, we expect similar results for liquid Xe. These results, if taken at face value, suggest that the detection of SUSY WIMPs could be very difficult, requiring large detectors of certainly 100 kg and possibly tons of detector. In this case, the rejection of background is even more important. Table 1 gives the schematic for WIMP detectors. Figures 4–6 show some detection methods that will be discussed below.

In order to discover WIMPs interacting in a medium, the response of the medium must be extremely well understood. Discoveries are not made by removing backgrounds but by identifying a unique signature for the process. Table 2 lists the important signature for a WIMP interaction in liquid xenon. This information builds on more than two decades of study of the excitations in liquid Xe. The UV scintillation light in xenon is produced by the formation of Excimer states, which are bound states of ion–atom systems (see Fig. 7). There are extensive studies in the use of this process for Excimer lasers, as well as for many other applications.[10-12]

A successful test of the detection of a recoil Xe nucleus using neutron scattering has been recently carried out, and it shows clear evidence that SUSY WIMPs will give a strong, unique signal on a discriminating liquid-Xe detector (Figs. 8–10). A 2-kg detector will be installed at the Mt. Blanc Underground Laboratory (UL) to perform a first search for SUSY WIMPs using this tchnique.

In Figs. 5–7, we show the schematic of several other methods of searching for WIMPs and some current results.

The 2-kg ICARUS–WIMPs detector has now been installed at the CNR Torino Mt. Blanc UL and first tests are underway (see Fig. 10). Our goal is to search for WIMPs to the level of 10^{-1} events/kg·d during the next year or so.[13]

CURRENT STATUS OF THE SEARCH FOR SUSY WIMPS

In the early days of the direct search for WIMPs, the motivation was the detection of a massive dirac neutrino, which is now excluded by the data (see Fig. 2). In this search, the expected rate was so large (~ 1000 events/kg·day) that the detector background was not extremely important. In the search for SUSY WIMPs, as was discussed earlier (Fig. 3), we expect much lower rates and either very massive detectors are needed (for the background limited search, which goes like $\sqrt{N}$) or powerful discrimination. The best current limits come from the study of the pulse-shape distribution in NaI or liquid Xe. One NaI detector (the UK group) is shown in Fig. 5, where the current best limits (from the Rome group) are also shown.

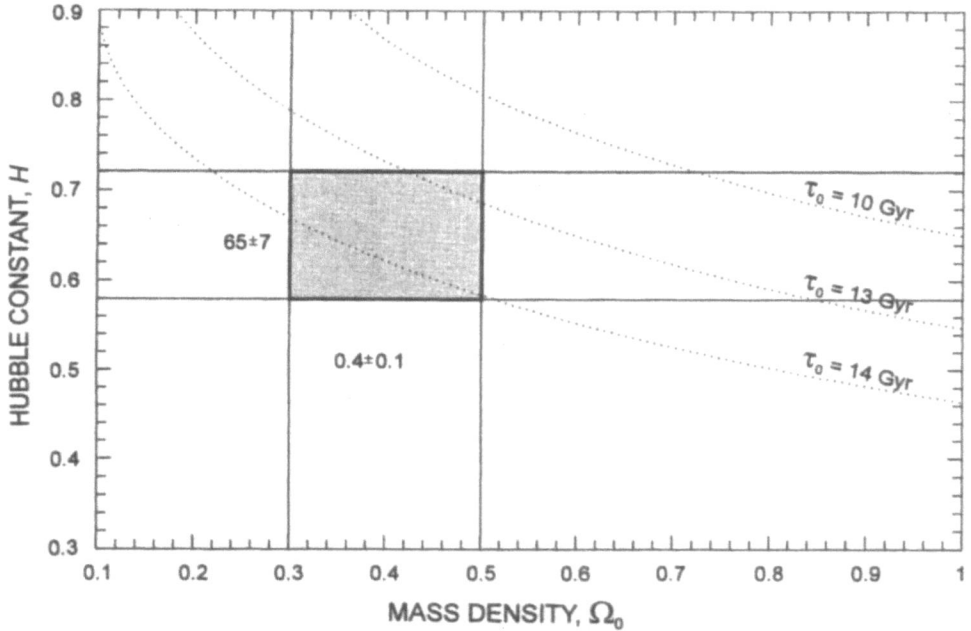

Figure 1. Recent estimates of the mass density as a function of Hubble's constant, and τ_0 is the age of the Universe.

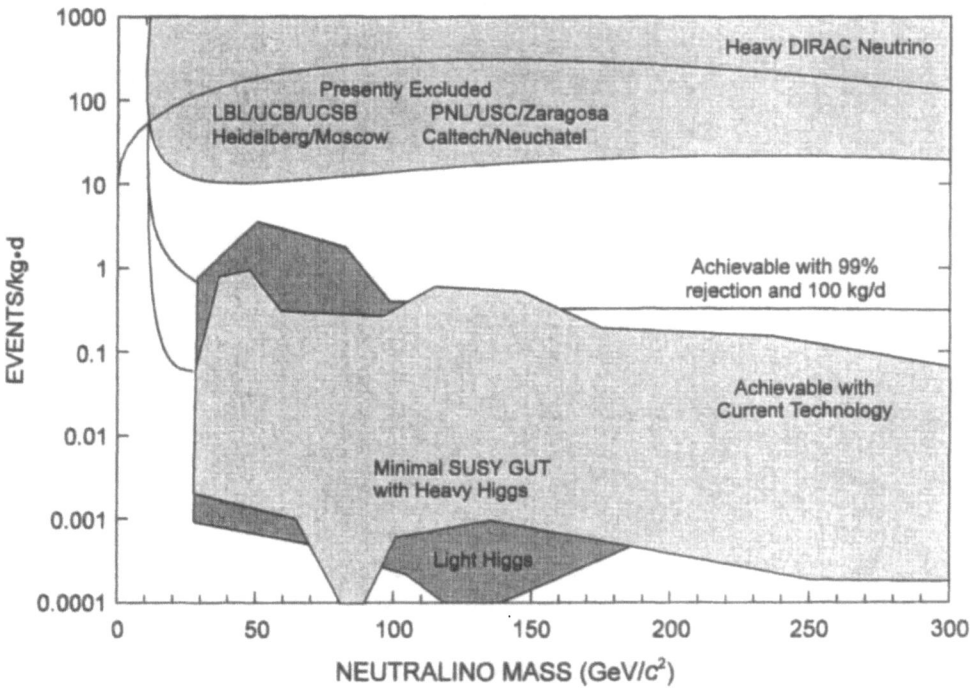

Figure 2. Regions of the event rate as a function of mass that were excluded in 1995 and the expected ranges for SUSY WIMPs.

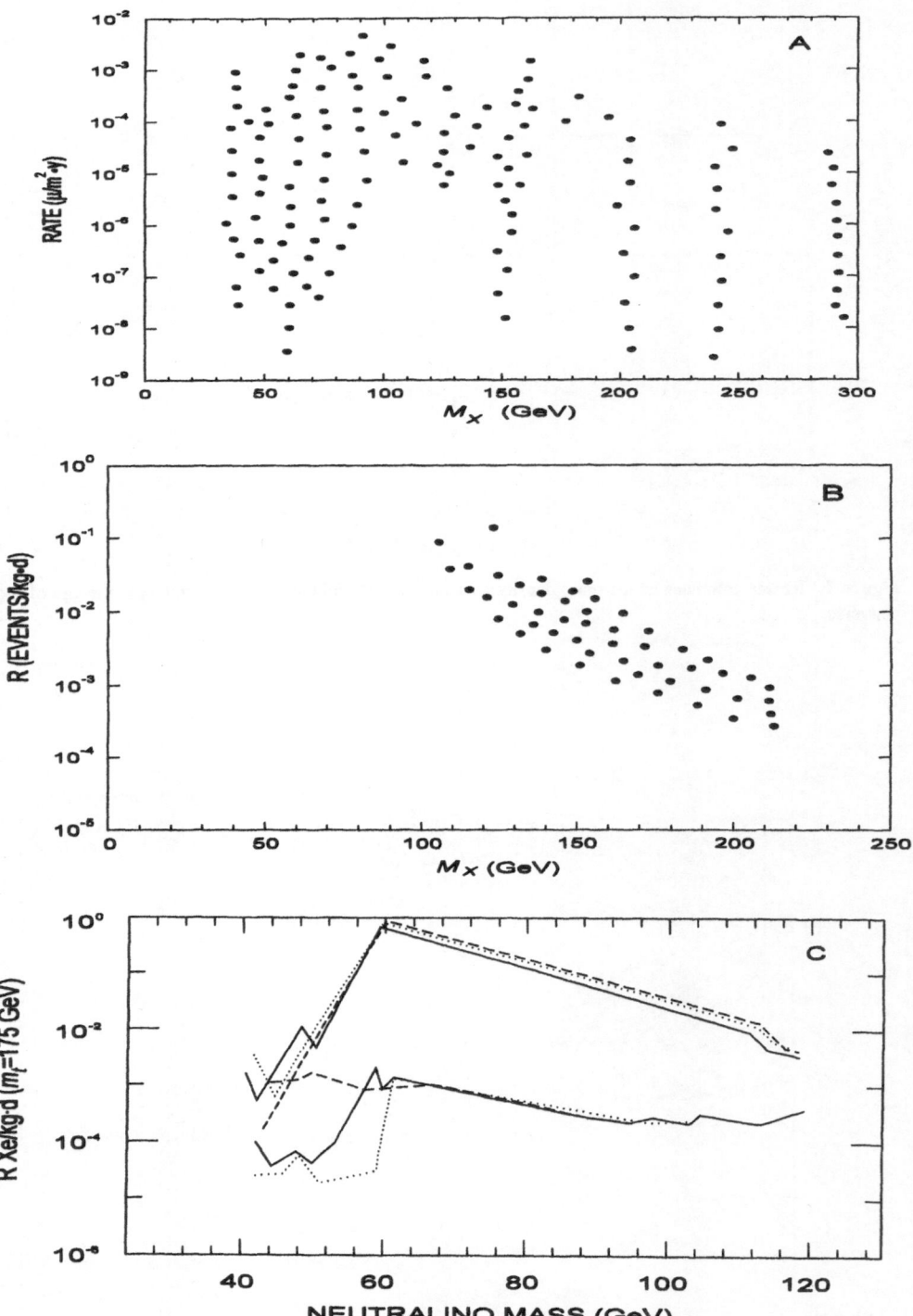

Figure 3. (A) Regions of the SUSY-WIMPs event rate (from Ref. 5); (B) Recent estimates of the events rate for SUSY WIMPs from the Gran Sasso CERN group (from Ref. 6); (C) Very recent estimates of the allowed regions of event rates (from Ref. 4).

TABLE 1. WIMP direct detection schematic.

T	T*

1. $W + [\text{atom/nucleus}] \rightarrow W + [\text{atom}]^* \sim$ elastic scattering with excitation of atom ~10 nm

2. Form factors for recoil [atom/nucleus] $\Delta E \leq 30$ keV measure $n + A \rightarrow n + A^*$

3. Kinematics

4. Rate $\leq (1 - 10^{-5}) \text{ kg}^{-1} \text{ d}^{-1}$

5. Background $- (10 - 100) \text{ kg}^{-1} \text{ d}^{-1}$

6. Discrimination against background

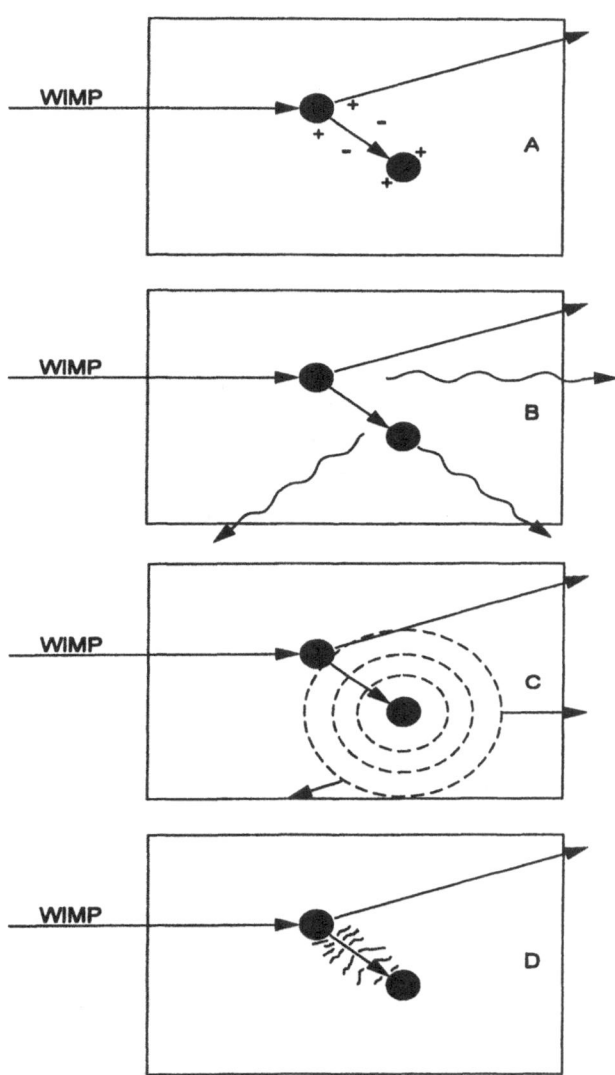

Figure 4. Schematic of WIMP detectors that use (A) semi-conductors, (B) scintillation light, (C) bolometers, and (D) etched mica.

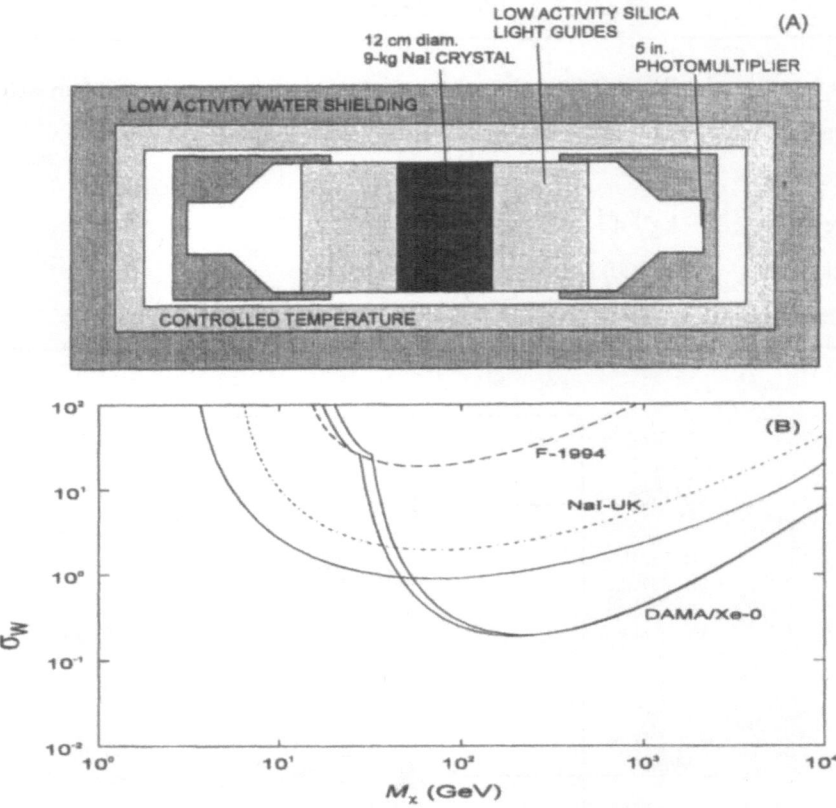

Figure 5. (A) The UK dark-matter-group NaI WIMP detector (from Ref. 7); (B) some recent limits on the WIMP cross section from NaI and Xe detectors (from Ref. 8).

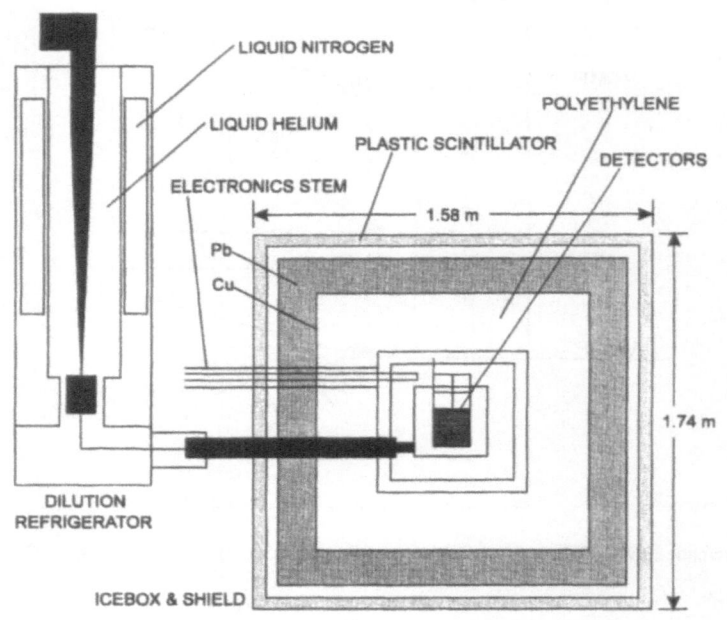

Figure 6. The CDMS detector at SLAC that uses the bolometer method (from Ref. 9).

Table 2. Signature and background in liquid xenon.

Recoil Nuclei
- Heavily ionizing particle
- High recombination, hence
- Mainly scintillation light is produced

Radioactivity
- Minimum ionizing particle
- Low recombination, hence
- Both charge and light are produced

In liquid Xe
- Both charge and light are visible
- This provides an efficient way for signal-to-background rejection

Moreover, in Xe
- No long-lived natural isotopes are present
- Xe^{127} has longest decay time (≈ 36 d)

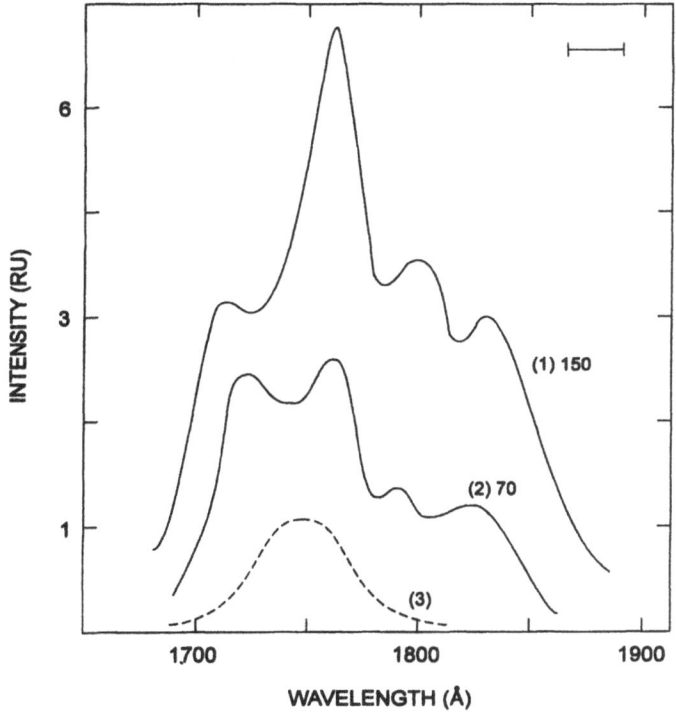

Figure 7. Emission spectrum of liquid Xe: (1 and 2) pump densities in Å/cm^2 and (3) emission spectrum at low excitation density. The resolution of the monochromator is shown in the upper right-hand corner. (From Ref. 10).

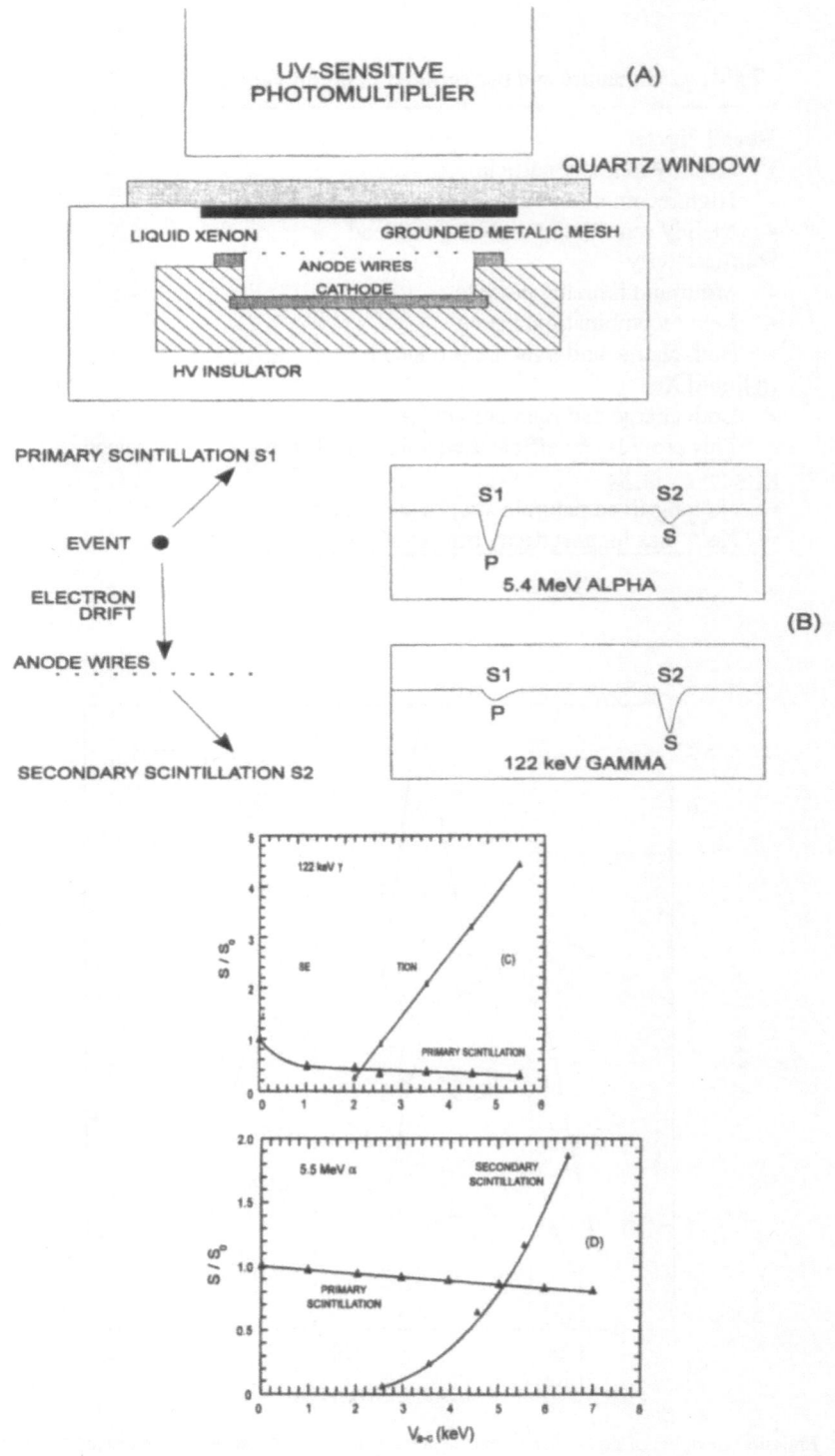

Figure 8. (A) Geometry of liquid-xenon test chamber, (B) observed primary and secondary scintillation signals showing S1/S2 >> 1 for α events and << 1 for γ events, (C) variations of the secondary scintillation intensity as a function of V_{a-c} for photons and (D) for α particles. (From Refs. 11–13.)

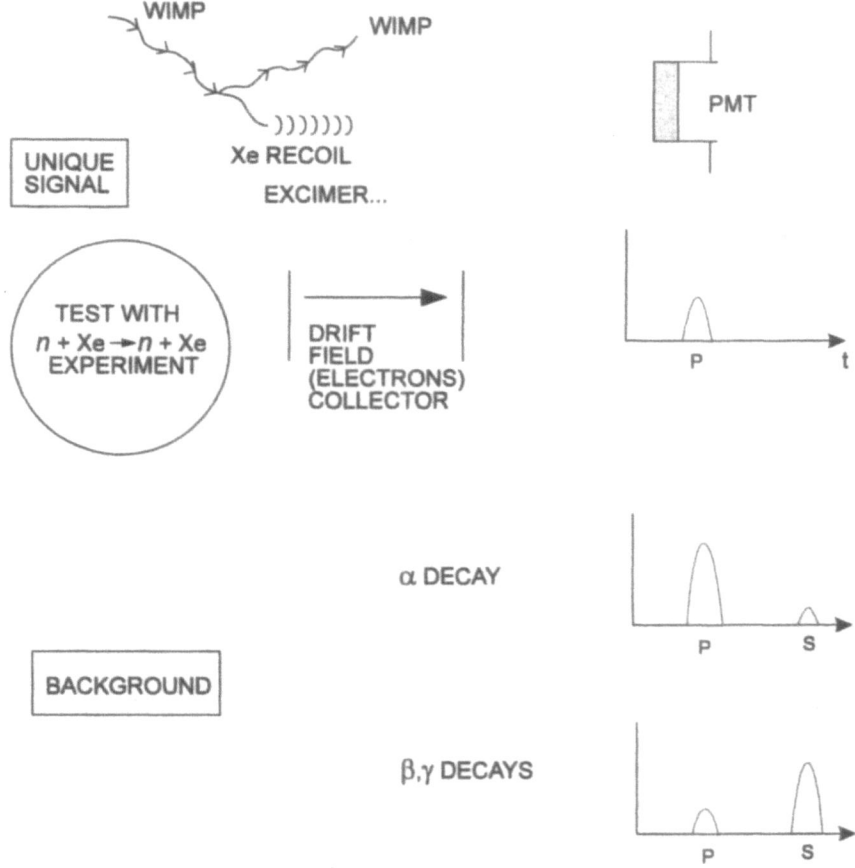

FIGURE 9. Schematic of the background rejection method in liquid Xe (rejection of 10^{-3}).

Discriminating Liquid-Xenon Detector[a]

To allow lower limits to be reached, it is essential to develop methods of differentiating the desired nuclear recoil events from γ- and β-decay backgrounds. At the same time, it is desirable to develop techniques capable of being substantially scaled up in target mass. The need for targets in the 100–1000-kg region would arise, particularly in searches for the 5% "annual modulation" of any true dark-matter signal (due to the Earth's motion combined with the solar motion through the Galaxy). Large-mass targets would also be needed for heavier WIMP masses (> 50 GeV), because of the correspondingly smaller flux of such particles.

Liquid Xe satisfies all of the above requirements for a dark matter detector because:
1. It is available in sufficiently large quantities with high purity.

2. It scintillates via two mechanisms, which are stimulated to different extents by nuclear-recoil and background electron-recoil events.

3. Its natural form consists of isotopes with and without nuclear spin, so it is suitable as a detector for both spin-independent and -dependent interactions.

The larger nuclear mass of Xe also makes it a better match to heavier WIMPs but, at the same time, the larger nuclear radius introduces a significant form-factor correction unless the energy

---

[a] This section is adapted from a recent report[12] by the ZEPLIN group.

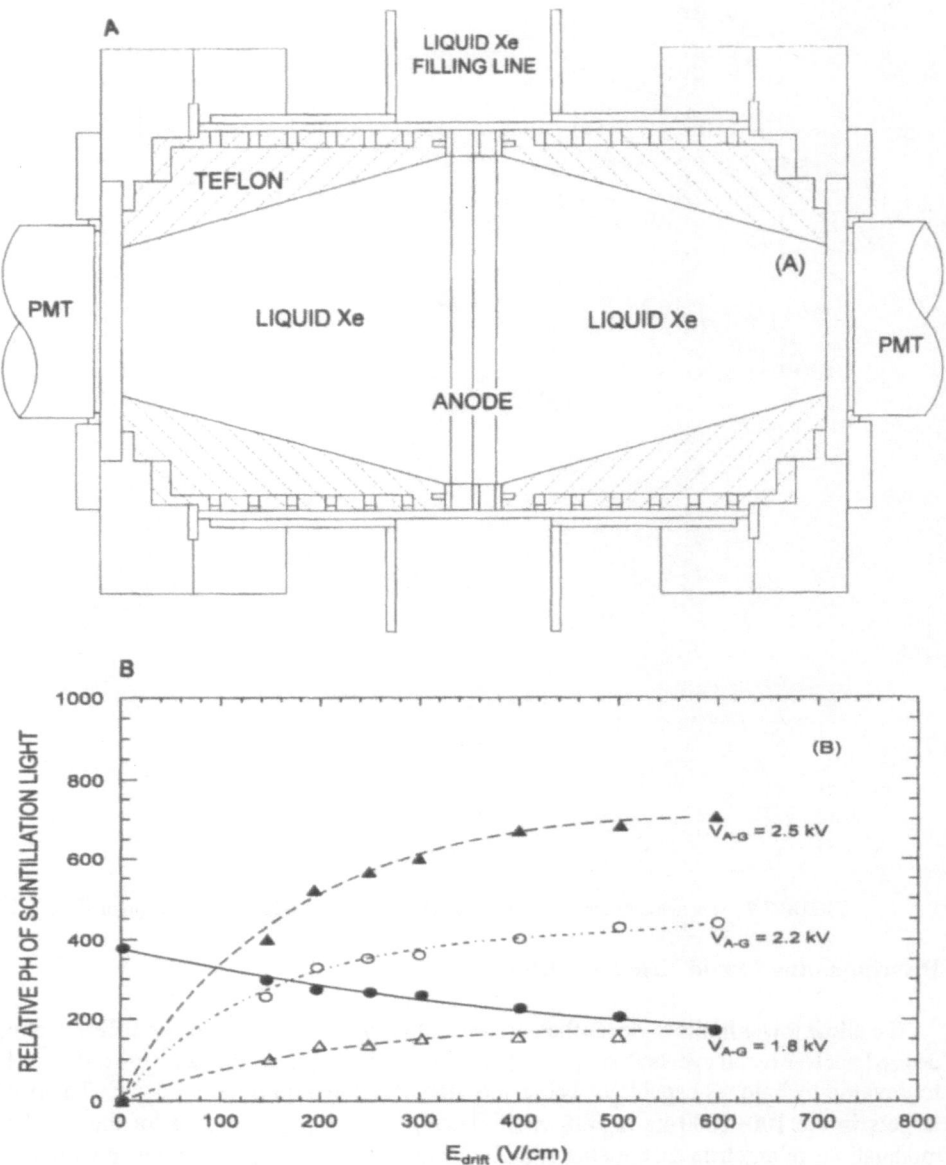

FIGURE 10. (A) A 2-kg detector that has been constructed for tests at Mt. Blanc and a possible WIMP search, and (B) variations of the secondary scintillation intensity as a function of E_{drift} and V_{A-G} for photons.

threshold is low (1–10 keV). Efficient light collection is, therefore, of prime importance in a liquid-Xe detector. Figure 11(A) shows a schematic of the proposed 20-kg ZEPLIN detector.

There are two distinct approaches to discriminating nuclear-recoil events in liquid xenon:

1. Analyzing the total scintillation pulse shape or, at low energy, the individual photon arrival times, which will differ significantly for nuclear- and electron-recoil events;

2. Applying an electric field to prevent recombination and measuring (i) the primary scintillation and (ii) the ionization component by drifting and producing "secondary scintillation" (see Figs. 7–8). The lowest possible backgrounds [Fig. 11(B)] may be achieved at the Boulby Mine site.

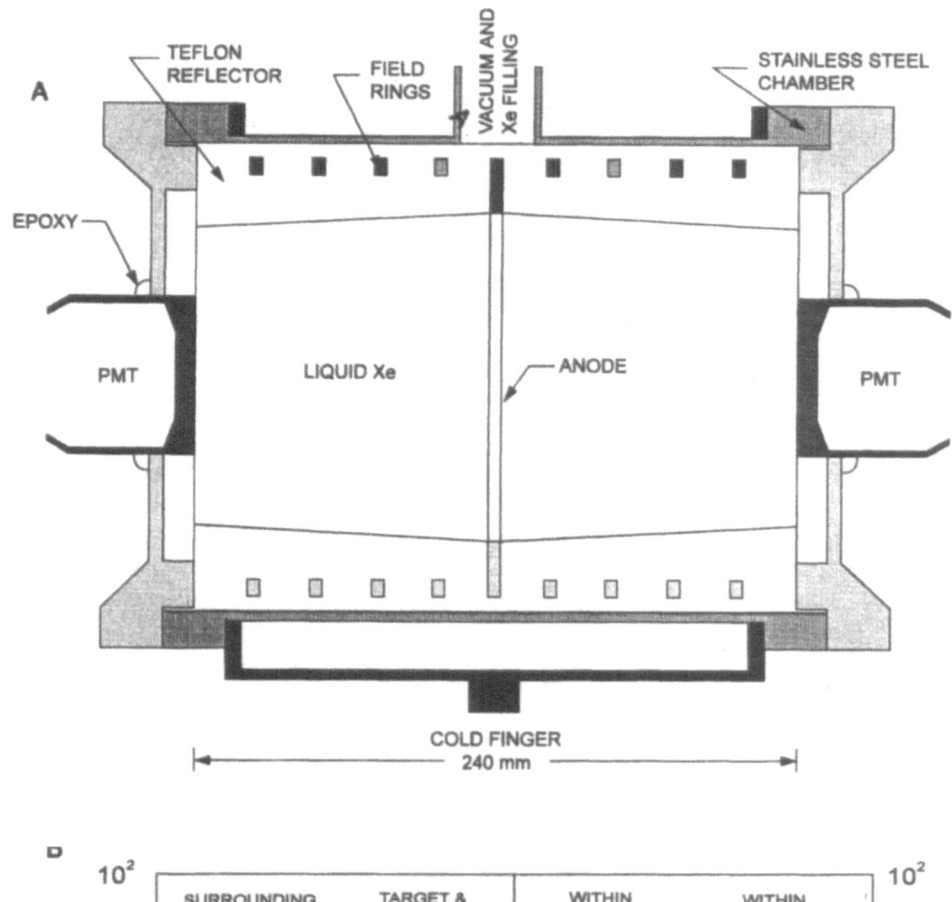

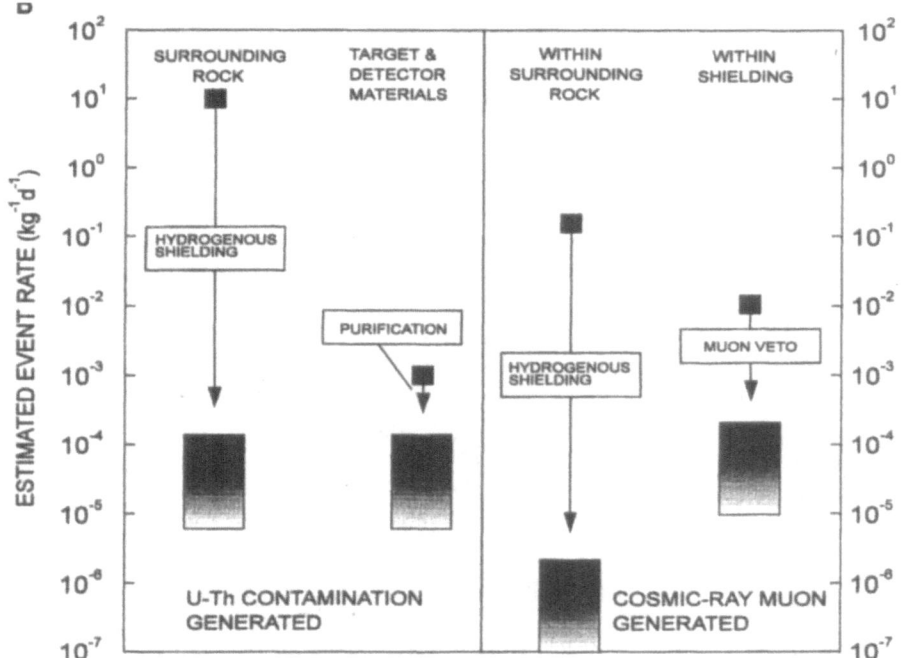

Figure 11. (A) Conceptual design for ZEPLIN system, showing inner proportional zone and outer shielding zone (total length, 40 cm), with PMT for collecting proportional scintillation light;[14] (B) Estimated background at the Boulby dark-matter laboratory in the UK.[15]

In Fig. 12 we show some estimates of how well the search for SUSY WIMPs can go using the Xe detectors discussed here.

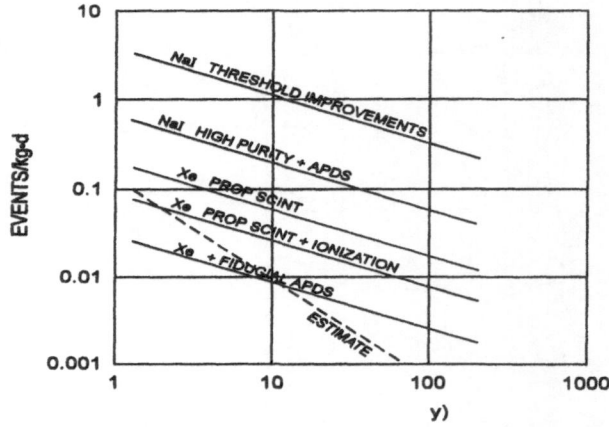

Figure 12. Possible sensitivity that can be reached with the Xe detetor discussed here (adapted from Ref. 15). The dashed lines show the sensitivity if the signal exceeds the background, as we expect for a discriminating liquid Xe detector.

SUMMARY

The search for SUSY WIMPs must now be one of the most important activities in particle physics. Recent estimates of the rate put it into the 10^{-1} - 10^{-5} event/kg·d level, which is an extremely difficult task. We believe the search is well worth the effort needed.

ACKNOWLEDGMENTS

I wish to thank Hanguo Wang for his excellent work on the ICARUS-WIMP detector and the other members of the ICARUS and ICARUS-WIMP groups (P. Picchi), as well as members of the proposed ZEPLIN group (P. Smith in particular); my thanks also to P. Nath and R. Arnowitt for discussions on the theory of SUSY-WIMP detection and for Fig. 3(C).

REFERENCES

1. See the proceedings of "Sources of Dark Matter in the Universe," Santa Monica, 1996, ed. D. Cline, *Nucl. Phys. B* (PS) 51B (1996); and L. H. Aller and V. Trimble, in *Sources of Dark Matter in the Universe*, ed. D. Cline (World Scientific, Singapore, 1994) pgs. 3 and 9.
2. F. Zwicky, *Ap. J.* 86:217 (1937).
3. CERN-UCLA-INFN Group, ICARUS Proposal (1993) unpublished.
4. P. Nath and R. Arnowitt, *Phys. Rev. Lett.* 74:4592 (1995).
5. G. Jungman *et al.*, *Phys. Reports* 267:195 (1996).
6. A. Bottino *et al.*, *Astropart. Phys.* 2:77 (1994).
7. P. Smith *et al.*, *Phys. Lett. B* 379:299 (1996).
8. R. Bernabei *et al.*, *Phys. Lett. B* 389:757 (1996).
9. B. Cabrera, *Nucl. Phys. B* (PS) 51B:297 (1996).
10. N. Basou *et al.*, *JETP Lett.* 12:329-331 (1970).
11. P. Benetti *et al.*, *Nucl. Instrum. Methods A* 327:203-206 (1993).
12. J. Park, M. Atac, D. B. Cline, H. Wang, and P. F. Smith, in *Sources of Dark Matter in the Universe*, ed. D. Cline (World Scientific, Singapore, 1994) p. 288.
13. H. Wang, CERN-UCLA, private communication (1996).
14. D. Cline, *Nucl. Phys. B* (PS) 51B:304-313 (1996).
15. P. Smith *et al*, *Nucl. Phys. B* (PS) 51B:284-293 (1996).

SEISMIC SEARCH FOR STRANGE QUARK MATTER

Eugene T. Herrin[1] and Vigdor L. Teplitz[2]

[1]Department of Geological Sciences
[2]Department of Physics
Southern Methodist University
Dallas, Texas 75275

ABSTRACT

We briefly review the status of strange quark matter and the capabilities of seismic networks for detecting nuggets of strange quark matter that pass through the Earth if an appreciable fraction of galactic halo dark matter is in the form of ton-sized nuggets. We describe progress to date in analyzing 13 years of seismic data and we note the possibility that new networks under development for verifying compliance with the recently negotiated Comprehensive Test Ban Treaty might be capable of contributing to detection of such nuggets.

INTRODUCTION

Witten[1] in 1984 pointed out that strange quark matter is more likely to be stable than non-strange quark matter since conversion to strange quarks (as long as $m_S < M_N/3$) lowers the Fermi energy. This fact was known to others[2]. Witten, however, went on to suggest that nuggets of strange quark matter could solve the cosmological dark matter problem by evading the bound on the baryon density from the deuterium abundance. He gave a scenario for producing nuggets in phase transitions in the early universe. Modified versions of his scenario are still under debate. He also noted that production could occur in supernovae and in neutron stars.

Farhi and Jaffe[3] considered in some detail the properties of nuggets of strange quark matter. Very briefly, they would have nuclear densities of 10^{14}gm/cm^3; there would be no saturation in baryon number A, but, instead, the binding energy per unit A would tend from below to a constant value, and the electron to baryon ratio would be on the order of 10^{-3} rather than $1/2$ as in ordinary nuclei. After their work, de Rujula and Glashow[4] considered terrestrial effects from incident strange quark nuggets (nuclearites in their terminology) with masses from nuclear to 10^9 gm where the maximum possible flux compatible with the local halo dark matter density drops below one per year. An impressive amount of additional study by many workers has gone into the area, and there is an excellent review of most of the basic ideas in the proceedings of the Arhus Workshop on Strange Quark Matter[5].

In his initial paper, Witten[1] raised the question of possible seismic detection of nuggets passing through the Earth. A nugget of mass much over a gram moving with galactic halo velocities will pass through the Earth, breaking the bonds between atoms in rock as it goes. De Rujula and Glashow[4] estimated the range of masses sufficient to produce a seismic signal strong enough for detection and outlined the differences between the pattern of seismic signals from an Earthquake, which is a point event, and that from nuclearite passage in which the signal would emanate from an entire line since the nugget speed (a few hundred kilometers per second) will greatly exceed that of seismic waves (around 10km/s).

We have pursued the issue of seismic detection[6]. This contribution will review progress to date. First, we have performed a Monte Carlo calculation based on a simplified model of the Earth, but real seismic stations, in order to estimate the Earth's sensitivity to nuclearite passage in practice. Second, we have accumulated 13 years of seismic reports and begun the process of analyzing them for nuclearite candidates. Finally, we should like to utilize this opportunity to suggest a further area of study: new networks currently being installed for the purpose of monitoring compliance with the recently-negotiated Comprehensive Test Ban Treaty might have a capability for detecting nuclearite passages, or for confirming future seismic detection.

WHAT MASS NUCLEARITES CAN CURRENT SEISMIC STATIONS DETECT?

In order to answer this question, we generated 120,000 nuclearite geometeries. To specify a nuclearite geometry requires six variables: two for the point at which it enters the Earth; three for the velocity vector; and one for the time of entry. We generated nuclearites so as to be isotropic in the rest system of the galaxy and then transformed to Earth-based geographical coordinates. This permitted us to see the effects of the motion of the sun and the Earth as in the work of Druckier, Freese and Spergel[7]. As nuclearites pass throughout the Earth their energy loss is given by the familiar formula.

$$d E/dx = -A\rho\upsilon \qquad (1)$$

where $\bar{\upsilon}$ and A are the nuclearite's velocity and area while ρ is the density of the Earth. We used a simplified model of the Earth, through which the seismic signal generated must propagate: spherical; uniform density; and single sound speed value. We assumed that five percent of the energy loss would go into the seismic signal. In comparison, the coupling for nuclear explosions is about 1%, and that for chemical explosions is about 2%. Nuclearite passage would appear to minimize the ratio of heat to signal, but we cannot rule out significant deviations from the 5% assumed.

We used a set of 278 seismic stations drawn from stations in existence during the 1980's. The best of these stations, of which there were 48, were assumed capable of detecting the signal from one kiloton of TNT at a distance of 5000 km. This corresponds to detecting a ground motion of 0.5-1.0 nm or 0.133 erg cm^{-2}s^{-1}. We computed signal propagation just as Jackson computes Cherenkov radiation[8]. If the nuclearite enters the Earth, at time t=0, at the origin, and with velocity υ and exits at time t_1, then its trajectory is $r = \vec{v}t'$. We can then write for the time t at which a signal is received at a station less the time t' at which it was emitted along the nuclearite line,

$$(t-t')_{\pm} = \left(\upsilon^2 - c^2\right)^{-1}\{(\vec{r}-\vec{\upsilon}t)\cdot\vec{\upsilon}\pm[(\vec{\upsilon}\cdot(\vec{r}-\vec{\upsilon}t))^2 - (\upsilon^2 - c^2)(\vec{r}-\vec{\upsilon}t)^2]^{\frac{1}{2}}\} \qquad (2)$$

where r is the location of the station and c is the velocity of sound in the simplified Earth model. The time at which the signal first arrives at the station is just given by the time t for which the square bracket in the Equation (2) vanishes.

$$t_F = \frac{\vec{\upsilon}\cdot\vec{r}}{\upsilon^2} \pm \frac{1}{\upsilon c}(1-\frac{c^2}{\upsilon^2})^{\frac{1}{2}}[r^2\upsilon^2 - (\vec{r}\cdot\vec{\upsilon})^2]^{\frac{1}{2}} \qquad (3)$$

This gives

$$t'_F = t_F + \vec{\upsilon}\cdot(\vec{r}-\vec{\upsilon}t)/(\upsilon^2 - c^2) \qquad (4)$$

There are thus three cases. For $t' < 0$ and $t' > t_1$, we have that the signal travels from the

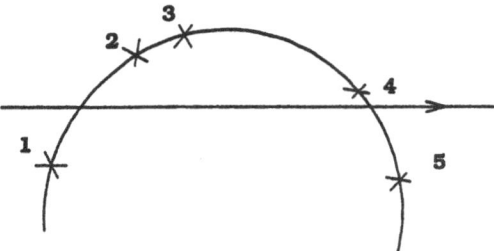

Figure 1. Difference between point and line events. An earthquake at time zero at the entry point would give arrival times $t_1 < t_2 < t_3$ and no signal from 4 and 5. A nuclearite would give $t_4 < t_2 < t_1 < t_3 < t_5$.

entry and exit point and t_F is given by r/c and $\dfrac{|\vec{r}-\vec{\upsilon}t_1|}{c}$ respectively. For times t' between these two limits, Equation (3) gives the signal arrival time. The first two cases correspond to signal arrivals at stations 1 and 5 in Figure 1. Figure 1 illustrates the difference between signal arrival patterns for line and point events.

Reference (6) gives the needed expressions for the intensity of the signal at each station. For each of our 120,000 geometries, we asked for the minimum mass that would provide a signal strong enough to be detected by seven stations. Our reasoning was that six station reports are sufficient to determine the nuclearite trajectory. If the same trajectory results from all seven determinations that can be made with seven reports, that set of reports would be recognized as a strong candidate. Our results were that one third of nuclearite geometries require masses under ten tons and 79% require masses under 100 tons. These correspond to nuclearite diameters of a few tens of microns. Taking account of the fact that, if there is a given amount of mass to be split up you can get a larger number of small-mass nuclearites, the relative detection probability peaks at about 11% around four tons and is near 10% in the range 1-10 tons. Figure 2 shows the limit on the ratio R of mass density in nuggets to total estimated halo dark matter mass density from a bound of less than one event detected per year. Finally, in Figure 3, we show a projection of the 48 most sensitive stations and an approximation to a nuclearite trajectory (the entry and exit points are joined simply by a straight line) with the signal arrival times ordered; dramatic videotapes of signal arrivals were shown at the oral presentation.

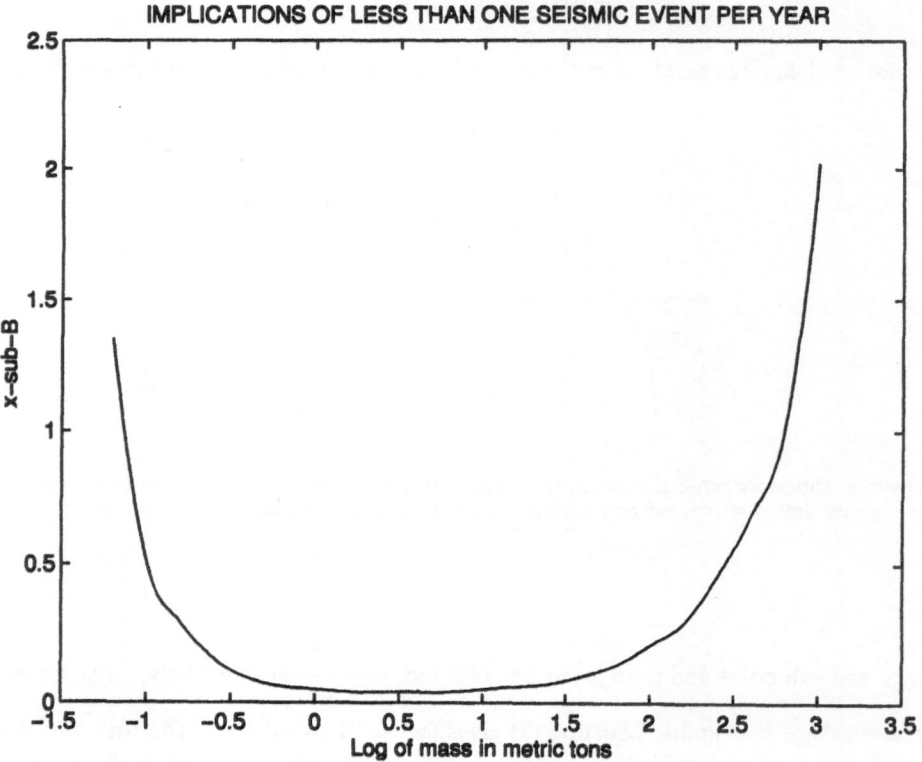

Figure 2. Limit on x_B (ratio of mass density in SQM nuggets of mass M to estimated halo DM density) from a bound of less than one event detected per year.

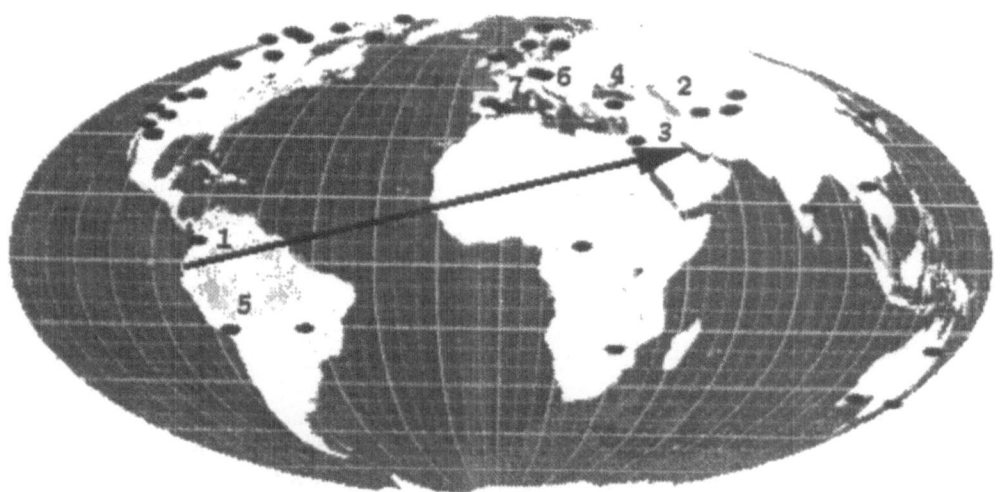

Figure 3. The 48 most sensitive stations and one Monte Carlo nuclearite event with entry and exit points joined by straight line and with stations numbered in order of receipt of signal.

REAL DATA

Reports of seismic signals are sent by many stations to the U.S. Geological Survey (USGS). They have graciously permitted us to copy their reports for the years 1981-1993[9]. There are nine million reports in the record; six million have been associated into earthquakes. It is then from the three million unassociated reports that we seek to find nuclearite candidates. It should be emphasized that the reports from seismic stations have all be reviewed by trained personnel before being forwarded to USGS. The average time between unassociated reports is on the order of three minutes while the time for sound or seismic waves to cross the Earth is about ten. Thus the chances of getting seven or more random reports in ten minutes are not small. We have about 100,000 such cases. We have investigated a dozen or so sets of reports of large numbers of signals within ten minutes and found that, in most cases, the reports are ones that should have been associated with Earthquakes but were not. Our plan is to make cuts on the data by removing all reports from the data set if their seismic activity occurred within an hour of an earthquake over some given value in magnitude, and to adjust the value so as to minimize the background while leaving as much time in which signal can occur as possible.

A second effort needed for the search is to replace the simple Earth model used for the Monte Carlo calculation with a more realistic one that takes into account the variation with depth below the earth's surface of the speed of sound. Because of the variation, seismic waves travel a path curved in such a way as to minimize travel time. Travel times between points in the Earth (event sources) and stations on the surface have only been calculated for the case of points in the Earth within 750 km of the surface since it is in this region that all earthquakes are believed to occur. A talented student is calculating travel times for signals originating at arbitrary points in the Earth's mantle down to the interface with the core. Because of the complexity of travel through the core, we will not, at least initially, attempt to model signals originating from within it. We will use these travel times in looking for nuclearite geometries that yield observed values of signal arrival times for sets of seven or more events within ten minutes. Because there has been relatively little seismic analysis directed at looking for signals from more than 750 km below the surface, this work could end up in new seismic insights even if nuclearites are not found.

POSSIBLE USE OF CTBT VERIFICATION SYSTEMS

There are three new verification systems under construction in connection with the recently negotiated Comprehensive Test Ban Treaty (CTBT)[10]. They are: (1) a seismic network of 50 primary stations augmented by an auxiliary network of 119 additional, less sensitive stations; (2) a hydroacoustic network of six fixed cable hydrophone stations and five land-based T-phase stations; and (3) an infrasound network of 60 stations. These international resources are expected to have continuous operation with high reliability in view of the importance of their mission. Their data might be able to contribute to nuclearite detection. The work of reference (6) provides a basis for a judgment as to the comparable analysis of the detection potential of the hydroacoustic and infrasound networks. Such an analysis would clearly be desirable.

CONCLUSIONS

Although we know it is more likely to be stable at zero pressure than matter made from up and down quarks, we do not know whether or not strange quark matter is stable. If it is, we do not know whether strange quark nuggets can constitute a significant fraction of the dark matter to which this session is devoted. However, should that be the case, and should a significant amount be in the form of nuggets in the one to 100 ton range, we have shown that seismic detection should be feasible. We have an ongoing program to search 13 years of seismic data for evidence of such nuggets passing through the Earth. There may be additional possibilities for detection, not previously considered, in the hydroacoustic and infrasound networks currently being constructed for CTBT verification.

ACKNOWLEDGMENTS

We very much appreciate the continuing, highly creative assistance of David Anderson in performing the numerical computations and graphical simulations and that of Ileana Tibuleac in constructing the critical travel times tables discussed in the text. We also have benefited from interaction with John Broderick, Brian Dennison, Timothy Krisher, Doris Rosenbaum, and Mark Sher, with whom other work on strange quark matter is in progress.

REFERENCES

[1] E. Witten, *Phys. Rev.* D30, 279 (1984).

[2] See, for example, A.R. Bodmer, *Phys. Rev.* D 4, 1601 (1971); B.A. Freedman and L.D. McLerran, ibid. 16, 1130 (1977); 16,1147 (1977); 17, 1109 (1978).

[3] E. Farhi and R.L. Jaffe, *Phys. Rev.* D30 2379 (1984).

[4] A. De Rujula and S.L. Glashow, *Nature* (London) 312, 734 (1984).

[5] J. Madsen and P. Haensel (eds), *Nuc. Phys.* B (Proc. Suppl.) 24B, 1 (1991).

[6] E.T. Herrin and V, L. Teplitz, *Phys. Rev.* D 53, 6762 (1996).

[7] A.K. Krukier, K. Freese, and D.N. Spergel, *Phys. Rev.* D 33, 3495 (1986).

[8] J.D. Jackson, *Classical Electrodynmalcs,* 2nd ed., John Wiley, New York, (1975).

[9] We very much appreciate the U.S. Geological Survey providing access to these reports and the help of James N. Taggart in arranging that access.

[10] See http://www.cdidc.org:65120/ for a description of the CTBT sensor networks.

SECTION V
GAUGE SYMMETRIES, GRAVITY AND STRINGS

GAUGE SYMMETRY IN
FIVEBRANE CONFORMAL FIELD THEORY

L. Dolan

Department of Physics, University of North Carolina

Chapel Hill, North Carolina 27599-3255, USA

INTRODUCTION

String duality provides us with a definition of string theory for large string coupling constant[1,2]. The massless particles of the perturbative spectrum of string theory remain massless for all coupling constant values. However string duality shows there is an additional solitonic set of states, absent in string perturbation theory, which for strong coupling are generically also of low mass. (In special cases the soliton states are massless even for weak coupling constant values.) These additional degrees of freedom have been described in some cases by a world-sheet conformal field theory (cft). One example of this is the D-brane open and closed string perturbation theory whose spectrum is the same as that of closed type II superstrings with Ramond-Ramond solitons[3,4].

Conformal field theory descriptions of Neveu-Schwarz solitons are also known. In fact the presence in Type IIB ten-dimensional string theory of both Ramond and Neveu-Schwarz fivebrane solitons suggests that their world-sheet descriptions should be related since the Type IIB string is $SL(2, \mathbf{Z})$ S dual to itself[5,6]. The importance of new string soliton states including those corresponding to the Ramond threebrane which couples to the self-dual four form potential is to help guide us in selecting a realistic string vacuum. Clearly non-perturbative string physics enhances the particle spectrum.

In this talk we present the superconformal field theory corresponding to the semi-wormhole throat limit of the Neveu-Schwarz fivebrane[7,8] and calculate the non-zero $SU(2)$ charge of massless space-time fermions in a string model containing the fivebrane cft as internal degrees of freedom[9]. The world-sheet model involves a Liouville conformal field theory which has a background charge. The norm of the states is defined together with an inner product and hermiticty properties. The connection between these states and those of the continuum description of 2D quantum gravity coupled to $c = 1$ matter, ie 2D string theory[10] which has discrete states is discussed.

FIVEBRANE CONFORMAL FIELD THEORY

The solitonic fivebrane solution in sigma-model coordinates has a metric given by

$$ds^2 = \eta_{MN} dx^M dx^N + (1 + \frac{k_6}{y^2} e^{C_0}) \delta_{mn} dy^m dy^n \tag{1}$$

where $0 \leq M, N \leq 5$, and $6 \leq m, n \leq 9$. The first term corresponds to a $c = 9$ cft, and the second term represents a $c = 6$ cft, where k_6 and C_0 are certain constants which contain a parameter k, the axion charge. We note that the (M,N) space-time part of the metric is flat, so that the $c = 9$ part is a free conformal field theory.

The $c = 6$ part is a nontrivial conformal field theory, and we call it the $c = 6$ "fivebrane" conformal field theory. This conformal model is given by the following: It contains four affine Kac-Moody currents J^A of dimension one satisfying an $U(1) \times SU(2)$ KMA (of level k) together with a set of four dimension-$\frac{1}{2}$ fields ψ^A satisfying the free fermion algebra:

$$J^0(z)J^0(\zeta) = -(z - \zeta)^{-2} + \dots$$
$$J^0(z)J^i(\zeta) = O(z - \zeta)^0$$
$$J^i(z)J^j(\zeta) = -\frac{k}{2}(z - \zeta)^{-2} + \epsilon_{ijk}J^k(\zeta)(z - \zeta)^{-1} + \dots$$
$$\psi^A(z)\psi^B(\zeta) = -(z - \zeta)^{-1}\delta^{AB} + \dots$$
$$\psi^A(z)J^B(\zeta) = O(z - \zeta)^0. \tag{2}$$

The energy-momentum tensor from the Sugawara construction is:

$$\tilde{L}(z) = -\frac{1}{2}J^0 J^0 - \frac{1}{k+2}J^i J^i - \frac{1}{2}\partial\psi^A \psi^A + \delta\frac{1}{4z^2} \tag{3}$$

where $\delta = 0, 1$ in the NS or R sector respectively; and $A = (0, i)$. The supercurrent is

$$\tilde{F}(z) = \psi^0 J^0 + \frac{\sqrt{2}}{\sqrt{k+2}}\psi^i J^i + \frac{\sqrt{2}}{6\sqrt{k+2}}\epsilon_{ijk}\psi^i \psi^j \psi^k. \tag{3}$$

The superconformal system (3) has $\tilde{c} = \frac{6(k+1)}{(k+2)}$. In order to achieve the $c = 6$ system corresponding to the fivebrane, a shifted cft is defined from (3):

$$\bar{L}(z) = \tilde{L}(z) + \frac{1}{2}Q\partial J^0(z)$$
$$\bar{F}(z) = \tilde{F}(z) - Q\partial\psi^0(z). \tag{4}$$

The superconformal system (3) has $c = \tilde{c} + 3Q^2$. We choose the background charge $Q = -\frac{\sqrt{2}}{\sqrt{k+2}}$, so that $c = 6$.

For $\bar{L}(z)$ hermitian, i.e. $\bar{L}_n^\dagger = \bar{L}_{-n}$, then $J_n^{0\dagger} = -J_{-n}^0$ for $n \neq 0$ since Q is real, $J_0^{0\dagger} = -Q - J_0^0$ $\psi_n^{A\dagger} = -\psi_{-n}^A$, and $J_n^{i\dagger} = -J_{-n}^i$. Therefore the states $J_{-n}^0|\psi\rangle$ for $n > 0$ have positive norm:

$$\||J_{-n}^0|\psi\rangle\|^2 = \langle\psi|(-J_n^0)J_{-n}^0|\psi\rangle = n\langle\psi|\psi\rangle = n\||\psi\rangle\|^2 > 0 \tag{5}$$

for $\||\psi\rangle\|^2 > 0$. A similar argument holds for $\psi_{-n}^A|\psi\rangle$ and $J_{-n}^i|\psi\rangle$. It follows from the above hermiticity conditions on J_n^A, ψ_n^A that $\bar{F}_n^\dagger = \bar{F}_{-n}$.

In order to study vertex operators, we consider the primary weight one-half superfield whose components are

$$V_L^i(z) = \psi^i; \qquad V_U^i(z) = Q(\tfrac{1}{2}\epsilon_{ijk}\psi^j\psi^k + J^i) \equiv T^i, \tag{6}$$

where

$$\bar{F}(z)\psi^i(\zeta) = (z-\zeta)^{-1}T^i(\zeta)$$
$$\bar{F}(z)T^i(\zeta) = (z-\zeta)^{-2}\psi^i(\zeta) + (z-\zeta)^{-1}\partial\psi^i(\zeta). \tag{7}$$

We define $T^i \equiv V_U^i$ and (8) forms a representation of a super Kac-Moody algebra

$$T^i(z)T^j(\zeta) = -\frac{\delta_{ij}}{(z-\zeta)^2} + \frac{Q\epsilon_{ijk}T^k(\zeta)}{(z-\zeta)} \tag{8a}$$

$$T^i(z)\psi^j(\zeta) = \frac{Q\epsilon_{ijk}\psi^k(\zeta)}{(z-\zeta)} \tag{8b}$$

$$\psi^i(z)\psi^j(\zeta) = -\frac{\delta_{ij}}{(z-\zeta)}. \tag{8c}$$

The level of the KMA T^i is $k+2$. If we bosonize the J^0 current by $J^0(z) = -\partial\phi(z)$, where $\phi(z)\phi(\zeta) = -\ln(z-\zeta) + \dots$ for $|z| > |\zeta|$, then the conformal field $: e^{\beta\phi(z)} :$ with $: e^{\beta\phi(0)} : |0\rangle = |\beta\rangle$ is primary with respect to $\bar{L}(z)$ with conformal weight $h_Q = -\tfrac{1}{2}\beta(\beta+Q)$. The operator product

$$J^0(z) : e^{\beta\phi(\zeta)} := \beta : e^{\beta\phi(\zeta)} : (z-\zeta)^{-1} + \dots \tag{9}$$

implies $J_0^0|\beta\rangle = \beta|\beta\rangle$. The anomalous product

$$\bar{L}(z)J^0(\zeta) = Q(z-\zeta)^{-3} + J^0(\zeta)(z-\zeta)^{-2} + \partial J^0(\zeta)(z-\zeta)^{-1} + \dots \tag{10}$$

implies $[\bar{L}_n, J_m] = -mJ_{n+m} + \frac{Q}{2}n(n+1)\delta_{n,-m}$. For a cft with background charge, considering the moments of the current $J^0(z)$ and requiring $\bar{L}_n^\dagger = \bar{L}_{-n}$, we derive

$$J_0^{0\dagger} = -[\bar{L}_{-1}, J_1^0]^\dagger = -[\bar{L}_1, J_{-1}^0] = -J_0^0 - Q. \tag{11}$$

The allowed values for β such that h_Q is real are given below. The case we consider in this paper is real Q and real h_Q. For $h_Q \equiv -\tfrac{1}{2}\beta(\beta+Q)$ to be real, either $\beta = -\tfrac{1}{2}Q+iy$ with real y so that $h_Q = \tfrac{1}{8}Q^2 + \tfrac{1}{2}y^2$ is positive; or β is real. If β is real, the highest weight conditions for massless unitary representations require it to take on discrete values, a feature similar to the discrete states in $c=1$ matter coupled to $2d$ quantum gravity, i.e. the Liouville field.[10]

GAUGE SYMMETRY IN BACKGROUND CHARGE CFT

We consider compactification of the Type II superstring with four-dimensional Lorentz invariance. There is a $c=9$, $N=1$ internal (local) superconformal algebra (SCA) associated with any such compactification. For theories with $d=4$, $N=2$ spacetime supersymmetry, the spacetime supersymmetry implies that this internal SCA splits into a $c=6$ piece with $N=4$ superconformal (global) symmetry and a

$c = 3$ piece with $N = 2$ superconformal (global) symmetry. Here we use *local* to mean the BRST ghost system also satifies this algebra.

We consider the $M^{(4)} \otimes T^{(2)} \otimes W_k^{(4)}$ model, where $W_k^{(4)}$ is the $c = 6$ fivebrane cft given in (4), and $M^{(4)} \otimes T^{(2)}$ is the flat four-dimensional non-compact flat Minkowski spacetime times a two-dimensional compact flat torus, given in terms of six free superfields $M = (0 \leq \mu \leq 3; 4 \leq m \leq 5)$,

$$a^M(z)a^N(\zeta) = \eta^{MN}(z - \zeta)^{-2} + \cdots$$
$$\psi^M(z)\psi^N(\zeta) = \eta^{MN}(z - \zeta)^{-1} + \cdots$$
$$\psi^M(z)a^N(\zeta) = 0 \tag{12}$$

where one of the free superfields is chosen to be timelike, and the others to be Euclidean (spacelike); $\eta^{MN} = (-1, 1, 1, 1, 1, 1)$ and $a^M(z) \equiv i\partial X^M(z)$ where

$$X^M(z)X^N(\zeta) = -\eta^{MN} \ln(z - \zeta) + : X^M(z)X^N(\zeta) :$$
$$\psi^M(z)\psi^N(\zeta) = \eta^{MN}(z - \zeta)^{-1} + \cdots$$
$$\psi^M(z)a^N(\zeta) = O(z - \zeta)^0 . \tag{13}$$

Then

$$L(z) = \frac{1}{2} : a^M a_M : + \frac{1}{2}\partial\psi^M \psi_M + \delta\frac{3}{8z^2} + \bar{L}(z)$$
$$F(z) = a^M \psi_M + \bar{F}(z) \tag{14}$$

where $c = 15$.

The Ramond ground states of the $N = 1$ superconformal algebra are expressed in terms of spin fields S_A, $S_{\dot{A}}$. The Ramond fields satisfy

$$\bar{F}(z) S_A(\zeta)e^{\beta\phi(\zeta)} = (z - \zeta)^{-\frac{3}{2}}(\beta + Q)\frac{1}{\sqrt{2}}\gamma_A^{0\dot{B}}S_{\dot{B}}(\zeta)e^{\beta\phi(\zeta)}$$
$$+ O(z - \zeta)^{-\frac{1}{2}} \tag{15a}$$
$$\bar{F}(z) S_{\dot{A}}(\zeta)e^{\beta\phi(\zeta)} = (z - \zeta)^{-\frac{3}{2}}\beta\frac{1}{\sqrt{2}}\gamma_{\dot{A}}^{0B}S_B(\zeta) e^{\beta\phi(\zeta)}$$
$$+ O(z - \zeta)^{-\frac{1}{2}} , \tag{15b}$$

where $\gamma_A^{0\dot{B}} = \sigma^0$ and $\gamma_{\dot{A}}^{0B} = -\sigma^0$, so that $S_A(0)e^{-Q\phi(0)}|0\rangle$ and $S_{\dot{A}}(0)|0\rangle$ satisfy

$$\bar{F}_0 S_A(0)| - q\rangle = 0, \quad \bar{F}_0 S_{\dot{A}}(0)|0\rangle = 0 \tag{15c}$$
$$\bar{F}_0 S_A(0)|0\rangle = \frac{q}{\sqrt{2}}S_{\dot{A}}(0)|0\rangle, \quad \bar{F}_0 S_{\dot{A}}(0)| - q\rangle = \frac{q}{\sqrt{2}}S_A(0)| - q\rangle . \tag{15d}$$

Here we choose a Weyl representation for the four-dimensional γ matrix algebra with $\{\gamma^m, \gamma^n\} = -2\delta^{mn}$ given by

$$\gamma^0 = \begin{pmatrix} 0 & (-\sigma^0)^A_{\dot{B}} \\ (\sigma^0)^{\dot{A}}_B & 0 \end{pmatrix} ; \quad \gamma^i = i\begin{pmatrix} 0 & (\sigma^i)^A_{\dot{B}} \\ (\sigma^i)^{\dot{A}}_B & 0 \end{pmatrix} \tag{16a}$$

where $\sigma^0 \equiv \begin{pmatrix} 1 & 0 \\ 0 & 1 \end{pmatrix}$ and σ^i are the Pauli matrices. The charge conjugation matrices

$$C = \begin{pmatrix} (i\sigma^2)^{AB} & 0 \\ 0 & (i\sigma^2)^{\dot{A}\dot{B}} \end{pmatrix} ; \quad C^{-1} = \begin{pmatrix} (-i\sigma^2)_{AB} & 0 \\ 0 & (-i\sigma^2)_{\dot{A}\dot{B}} \end{pmatrix} \tag{16b}$$

raise and lower indices: $C_{AD}^{-1}(\gamma^m)^D{}_{\dot{B}} = (\gamma^m)_{AB}$ and $C^{\dot{B}\dot{D}}(\gamma^m)^A{}_{\dot{D}} = (\gamma^m)^{A\dot{B}}$, etc. In deriving (15), we use the operator product expansion

$$\psi^{(0,i)}(z)S_A(\zeta) = (z-\zeta)^{-\frac{1}{2}}\frac{1}{\sqrt{2}}\gamma_A^{(0,i)\dot{B}}S_{\dot{B}}(\zeta) + \dots$$

$$\psi^{(0,i)}(z)S_{\dot{A}}(\zeta) = (z-\zeta)^{-\frac{1}{2}}\frac{1}{\sqrt{2}}\gamma_{\dot{A}}^{(0,i)B}S_B(\zeta) + \dots$$

$$\psi^i(z)\psi^j(z)S_A(\zeta) = (z-\zeta)^{-1}\frac{1}{2}(\gamma^i\gamma^j)_A^B S_B(\zeta) + \dots, \text{ etc.} \quad (17)$$

and in this representation $\frac{1}{6}\epsilon_{ijk}(\gamma^i\gamma^j\gamma^k)_A^{\dot{B}} = -\gamma_A^{0\dot{B}}$; and $\frac{1}{6}\epsilon_{ijk}(\gamma^i\gamma^j\gamma^k)_{\dot{A}}^B = \gamma_{\dot{A}}^{0B}$.

In general, modular invariance requires certain projections of the spin fields which restrict the physical spectrum. In the direct product of the three conformal field theories: (spacetime cft $c = 6$) $\otimes$ (fivebrane cft $c = 6$) $\otimes$ (T^2 cft $c = 3$), we consider the BRST invariant weight one primary conformal fields for Ramond ground states in the superconformal ghost number $q = -\frac{1}{2}$ picture given by

$$V_{-\frac{1}{2}}^{(1)}(k,z) = u^{1\alpha}(k)S_\alpha(z)e^{ik_\mu X^\mu(z)}S_A(z)e^{-Q\phi(z)}S_n(z)e^{-\frac{1}{2}\varphi_{\mathrm{gh}}(z)} \quad (18a)$$

$$V_{-\frac{1}{2}}^{(2)}(k,z) = u^{2\dot{\alpha}}(k)S_{\dot{\alpha}}(z)e^{ik_\mu X^\mu(z)}S_A(z)e^{-Q\phi(z)}S_{\dot{n}}(z)e^{-\frac{1}{2}\varphi_{\mathrm{gh}}(z)} \quad (18b)$$

where $k_\mu k^\mu = 0$; S_α, $S_{\dot{\alpha}}$ are the spin fields of 4-dimensional spacetime, and S_n, $S_{\dot{n}}$, $n = 1, \dot{n} = \dot{1}$ are the weight $\frac{1}{8}$ spin fields associated with the fermionic component ψ^ℓ of the supercoordinates on T^2. In (18), the two linearly independent solutions to the massless Dirac equation $k\cdot\gamma u(k) = 0$ are given by solutions of the Weyl equations

$$k_\mu\sigma^{\mu\dot{\alpha}}{}_\beta u^{1\beta} = 0 \qquad k_\mu\bar{\sigma}^{\mu\alpha}{}_{\dot{\beta}}u^{2\dot{\beta}} = 0 \quad (19)$$

and $\bar{\sigma}^0 = -\sigma^0$, $\bar{\sigma}^i = \sigma^i$. Since

$$iT^i(z)S_A(\zeta) = (z-\zeta)^{-1}Q\frac{i}{4}\epsilon_{ijk}(\gamma^j\gamma^k)_A^B S_B(\zeta)$$

$$iT^i(z)S_{\dot{A}}(\zeta) = (z-\zeta)^{-1}Q\frac{i}{4}\epsilon_{ijk}(\gamma^j\gamma^k)_{\dot{A}}^{\dot{B}}S_{\dot{B}}(\zeta) \quad (20)$$

we have

$$iT_0^3 S_1(0)|-q\rangle = -Q\frac{1}{2}S_1(0)|-q\rangle, \qquad iT_0^3 S_2(0)|-q\rangle = Q\frac{1}{2}S_2(0)|-q\rangle;$$

$$iT_0^3 S_{\dot{1}}(0)|0\rangle = -Q\frac{1}{2}S_{\dot{1}}(0)|0\rangle, \qquad iT_0^3 S_{\dot{2}}(0)|0\rangle = Q\frac{1}{2}S_{\dot{2}}(0)|0\rangle, \quad (21)$$

so that $V_{-\frac{1}{2}}^{(1,2)}(k,0)|0\rangle$ each also correspond to a doublet under the $SU(2)$ algebra T_0^i. Therefore these fermion states carry charge under the $SU(2)$ gauge algebra.

We note that the superfield associated with $\psi^0(z)$ has lower and upper components $\psi^0(z)e^{-Q\phi(z)}$, $T^0(z) = -\frac{1}{Q}\partial_z e^{-Q\phi(z)}$ so that unless $Q = 0$ the current $T^0(z)$ is not the generator of a $U(1)$ Heisenberg algebra.

The set of consistent GSO projections for states such as (18) in a string theory with space-time perturbative unitarity is being investigated. We note here that another set of BRST invariant Ramond operators is given by

$$V'^{(1)}_{-\frac{1}{2}}(k,z) = [u^{1\alpha}(k)S_\alpha(z)S_A(z) - \frac{Q}{\sqrt{2}}v^{1\dot{\alpha}}(k)S_{\dot{\alpha}}(z)\gamma_A^{0\dot{A}}S_{\dot{A}}(z)]$$

$$\cdot S_n(z)\,e^{ik_\mu X^\mu(z)}e^{-\frac{1}{2}\varphi_{\mathrm{gh}}(z)} \quad (22)$$

with a similar expression for $V'^{(2)}_{-\frac{1}{2}}(k,z)$. Vertex operators of this form, associated with different projections from (18), give rise to non-abelian tree amplitudes of Ramond-Ramond states.[9,11] Here we defined two additional spinors $v^\ell(k)$ by $k\cdot\gamma v^\ell \sim u^\ell$, i.e.

$$\frac{1}{\sqrt{2}}k_\mu\bar{\sigma}^{\mu\alpha}{}_{\dot{\beta}}v^{1\dot{\beta}} = u^{1\alpha} \qquad \frac{1}{\sqrt{2}}k_\mu\sigma^{\mu\dot{\alpha}}{}_\beta v^{2\beta} = -u^{2\dot{\alpha}}. \qquad (23)$$

For simplicity, we could have considered $k = 0$, in which case $Q = -1$ and the currents J^i decouple from the system, so that the superconformal/super Kac-Moody algebra with $c = 6$ given in (4,7,8) become:

$$L(z) = -\frac{1}{2}J^0J^0 - \frac{1}{2}\partial\psi^A\psi^A + \delta\frac{1}{4z^2} - \frac{1}{2}\partial J^0$$

$$F(z) = \psi^0J^0 + \frac{1}{6}\epsilon_{ijk}\psi^i\psi^j\psi^k + \partial\psi^0$$

$$\psi^i(z), \qquad T^i(z) = -\frac{1}{2}\epsilon_{ijk}\psi^j\psi^k$$

$$F(z)\psi^i(\zeta) = (z-\zeta)^{-1}T^i(\zeta), \quad F(z)T^i(\zeta) = (z-\zeta)^{-2}\psi^i(\zeta) + (z-\zeta)^{-1}\partial\psi^i(\zeta)$$

$$T^i(z)T^j(\zeta) = -\frac{\delta_{ij}}{(z-\zeta)^2} - \frac{\epsilon_{ijk}T^k(\zeta)}{(z-\zeta)}$$

$$T^i(z)\psi^j(\zeta) = \frac{\epsilon_{ijk}\psi^k(\zeta)}{(z-\zeta)}, \qquad \psi^i(z)\psi^j(\zeta) = -\frac{\delta_{ij}}{(z-\zeta)}. \qquad (24)$$

In (24), the degrees of freedom correspond to four fermions and one Liouville boson even though $c = 6$, a feature of non-critical string theory.

HERMITICITY PROPERTIES, INNER PRODUCTS AND NORMS

We consider the Liouville part of the cft described in (4) given by $L(z) = -\frac{1}{2} : J(z)J(z) : +\frac{1}{2}Q\partial J(z)$ so that for $n \neq 0$,

$$L_n = a_0a_n - J_0J_n - \frac{Q}{2}(n+1)J_n + \frac{1}{2}\sum_{m\neq 0, m\neq n} : a_{n-m}a_m : -\frac{1}{2}\sum_{m\neq 0, m\neq n} : J_{n-m}J_m : \qquad (25a)$$

and

$$L_0 = \frac{1}{2}a_0^2 - \frac{1}{2}J_0^2 - \frac{Q}{2}J_0 + \sum_{m=1}^\infty : a_{-m}a_m : - \sum_{m=1}^\infty : J_{-m}J_m : \qquad (25b)$$

The current $J(z)$ is anomalous:

$$L(z)J(\zeta) = Q(z-\zeta)^{-3} + J(\zeta)(z-\zeta)^{-2} + \partial J(\zeta)(z-\zeta)^{-1} + \dots$$

so that

$$[L_n, J_m] = \frac{Q}{2}n(n+1)\delta_{n,-m} - mJ_{n+m}$$

$$[L_{-1}, J_1] = -J_0$$

$$[L_1, J_{-1}] = Q + J_0 \qquad (26)$$

The primary field $e^{\beta\phi(z)}$ satisfies

$$L(z)e^{\beta\phi(\zeta)} = (z-\zeta)^{-2}[\frac{1}{2}p^2 - \frac{1}{2}\beta(\beta+Q)]e^{\beta\phi(\zeta)} + (z-\zeta)^{-1}\partial_\zeta e^{\beta\phi(\zeta)}$$

$$J(z)e^{\beta\phi(\zeta)} = (z-\zeta)^{-1}\beta e^{\beta\phi(\zeta)}$$

$$e^{\beta\phi(0)}|0\rangle = \psi_\beta$$

$$J_0\psi_\beta = \beta\psi_\beta \qquad (27)$$

where for $n > 0$, $J_n|0\rangle = 0$, $L_n|0\rangle = 0$.

As noted above, we can see from (25a) that for $L_n^\dagger = L_{-n}$, we must have that for $n \neq 0$, $J_n^\dagger = -J_{-n}$ if Q is real. From (26) it follows that $J_0^\dagger = -Q - J_0$.

These hermiticity properties correspond to the following definition of inner product:

$$(\psi_\mu, \psi_q) = \delta_{\mu, -q-Q} \tag{28}$$

since we must have

$$
\begin{aligned}
(\psi_\mu, J_0\psi_q) &= (J_0^\dagger \psi_\mu, \psi_q) \\
&= q(\psi_\mu, \psi_q) = ((-J_0 - Q)\psi_\mu, \psi_q) \\
&= (-\mu - Q)(\psi_\mu, \psi_q)
\end{aligned}
\tag{29}
$$

so that $(\psi_\mu, \psi_q) = 0$ unless $q = -\mu - Q$, hence (28).

We can write $(\psi_\mu, \psi_q) \equiv \langle \mu|q\rangle$ so that (28) is

$$(\psi_\mu, \psi_q) = \langle \mu|q\rangle = \delta_{\mu, -q-Q} .$$

For this definition of hermiticity and corresponding inner product, we have that the adjoint of $|q\rangle$ is $\langle q|$ and that

$$(\psi_q, \psi_q) = \delta_{q, -q-Q} = \delta_{q, -\frac{Q}{2}} \tag{30}$$

i.e. $\psi_q \equiv |q\rangle$ has zero norm unless $q = -\frac{Q}{2}$.

Note however that $|q\rangle$ is not null, ie it is not orthogonal to every state, since $(\psi_{-q-Q}, \psi_q) = \delta_{q,q} = 1 \neq 0$.

If the eigenvalues of J_0 are complex, then the RHS of (29) is $(-\mu^* - Q)(\psi_\mu, \psi_q)$, so that (30) is

$$(\psi_q, \psi_q) = \delta_{q, -q^*-Q} = \delta_{\mathrm{Re}q, -\frac{Q}{2}}$$

i.e. $\psi_q \equiv |q\rangle$ has **zero norm** unless $\mathrm{Re}q = -\frac{Q}{2}$.

Of course we could construct a positive norm state

$$\psi = \tfrac{1}{\sqrt{2}}\left(|q\rangle + |-q-Q\rangle\right) \tag{31a}$$

where $||\psi||^2 = 1$. But then there would also be negative norm state

$$\psi' = \tfrac{1}{\sqrt{2}}\left(|q\rangle - |-q-Q\rangle\right) \tag{31b}$$

where $||\psi'||^2 = -1$, unless this were absent by projection.

In the previous section, the norm of the spin field states $|\psi\rangle \equiv S_A(0)|0\rangle$ without the Liouville contribution is given by

$$
\begin{aligned}
\langle \psi|\psi\rangle &= \lim_{\substack{z \to \infty \\ \zeta \to 0}} z^{2h}\langle 0|V(\bar\psi, z)V(\psi, \zeta)|0\rangle = \lim_{\substack{z \to \infty \\ \zeta \to 0}} z^{\frac{1}{2}}\langle 0|S^A(z)S_A(\zeta)|0\rangle \\
&= \lim_{\substack{z \to \infty \\ \zeta \to 0}} z^{\frac{1}{2}}(z - \zeta)^{-\frac{1}{2}}C^{AB}C_{BA}^{-1} = \delta^A_A ,
\end{aligned}
\tag{32}
$$

161

where the vertex for the state $\bar{\psi}$ conjugate to ψ is given by $V(\bar{\psi}, z) \equiv z^{-2h} V(\psi, \frac{1}{z^*})^\dagger$. Since $S_{An}^\dagger = S_{-n}^A$, the state conjugate to $S_A(0)|0\rangle$ is $S^A(0)|0\rangle$. We have

$$\psi = \lim_{z\to 0} V(\psi, z)|0\rangle, \qquad \bar{\psi} = \lim_{z\to 0} V(\bar{\psi}, z)|0\rangle$$

$$\langle\bar{\psi}| = \lim_{z\to\infty} z^{2h}\langle 0|V(\psi, z), \qquad \langle\psi| = \lim_{z\to\infty} z^{2h}\langle 0|V(\bar{\psi}, z). \tag{33}$$

Therefore from (18) and (22), a Ramond state defined by

$$|\chi\rangle = V_{-\frac{1}{2}}^{(1)}(k, 0)\,|0\rangle + V_{-\frac{1}{2}}'^{(1)}(k, 0))\,|0\rangle \tag{34}$$

has positive norm and $SU(2)$ charge and is BRST invariant.

Let's now consider a **new** inner product

$$\langle\langle\psi_\mu, \psi_q\rangle\rangle = \delta_{\mu,q}. \tag{35}$$

This corresponds to the hermiticity properties $J_0^\dagger = J_0$, and for $n \neq 0$, $J_n^\dagger = -J_{-n}$ if Q is real, which results in $L_n^\dagger = L_{-n} + (2J_0 + Q)J_{-n}$ for $n \neq 0$, and $L_0^\dagger = L_0$. .

To show $J_0^\dagger = J_0$, we have[12] from (35) that

$$\langle\langle\psi_\mu, J_0\psi_q\rangle\rangle = \langle\langle J_0^\dagger\psi_\mu, \psi_q\rangle\rangle \tag{36a}$$

$$= q\langle\langle\psi_\mu, \psi_q\rangle\rangle = \mu\langle\langle\psi_\mu, \psi_q\rangle\rangle \tag{36b}$$

$$= \langle\langle J_0\psi_\mu, \psi_q\rangle\rangle \tag{36c}$$

where (36b) follows from $\langle\langle\psi_\mu, \psi_q\rangle\rangle = \delta_{\mu,q}$, and $J_0^\dagger = J_0$ follows from (36a,c).

We can write

$$\langle\langle\psi_\mu, \psi_q\rangle\rangle \equiv (\psi_{-\mu-Q}, \psi_q) = \delta_{\mu,q}$$

$$\langle\langle\psi_\mu, \psi_q\rangle\rangle \equiv \langle -\mu - Q|q\rangle = \delta_{\mu,q}. \tag{37}$$

For this definition of hermiticity and corresponding inner product, we have that the adjoint of $|q\rangle$ is $\langle -q - Q|$ and that

$$(\psi_q, \psi_q) = \delta_{q,q} = 1, \tag{38}$$

i.e. $\psi_q \equiv |q\rangle$ has unit norm.

If the eigenvalues of J_0 are complex, then the RHS of (36b) is $\mu^*\langle\langle\psi_\mu, \psi_q\rangle\rangle$, so that (38) is

$$(\psi_q, \psi_q) = \delta_{q^*,q} = 1,$$

i.e. $\psi_q \equiv |q\rangle$ has **non-zero norm** for any real q.

CONCLUSIONS

The specific fivebrane cft permits us to compute the spectrum of string excitations around the corresponding background solution. In this sense, the choice of cft provides non-perturbative information.

In this talk, we were led to consider a conformal field theory with background charge in order to have charged Ramond states. The fivebrane conformal field theory provides a specific example of such a conformal field theory, and its use was motivated by the fact that the fivebrane, an NS-NS brane, carries charge under the NS gauge bosons.

Furthermore, a connection between gauged supergravity theories which necessarily have tree level cosmological constants, and string theories serves to increase interest in pursuing non-critical string theories. The latter provide a link between string theory and field theories with tree level cosmological constants.

REFERENCES

1. C. Hull and P. Townsend, *Unity of superstring dualities*, Nucl. Phys. **B438** 109 (1995), hep-th/9410167; *Enhanced gauge symmetries in superstring theories*, Nucl. Phys. **B451** 525 (1995), hep-th/9505073.

2. E. Witten, *String theory in various dimensions*, Nucl. Phys. **B443** 85 (1995), hep-th/9503124.

3. J. Polchinski, *Dirichlet-branes and Ramond-Ramond charges*, Phys. Rev. Lett. **75** (1995) 4724, hep-th/9510017.

4. J. Dai, R. Leigh, and J. Polchinski, Mod. Phys. Lett. **A4** (1989) 2073.

5. J. Schwarz, *An SL(2,Z) multiplet of Type IIB superstrings*, Phys. Lett. **B360** (1995) 13; (E) **B364** (1995) 252, hep-th/9508143.

6. J. Schwarz, *Lectures on superstring and M-theory dualities*, hep-th/9607201.

7. A. Strominger, *Heterotic Solitons*, Nucl. Phys. **B343** (1990) 167.

8. C. Callan, J. Harvey, and A. Strominger, *Worldsheet approach to heterotic instantons and solitons*, Nucl. Phys. **B359** (1991) 611; and *Worldbrane actions for string solitons*, Nucl. Phys. **B 367** (1991) 60.

9. L. Dolan, *Gauge symmetry in background charge conformal field theory*, hep-th/9610091.

10. I. Klebanov and A. Polyakov, *Interaction of discrete states in two-dimensional string theory*, Mod. Phys. Lett A6(1991)3273 [hep-th/9109032]; I. Klebanov and A Pasquinucci, *Infinite symmetry and Ward identities in two-dimensional string theory*, hep-th/9210105; A. Polyakov, *Self-tuning fields and resonant correlators in 2d-gravity*, Mod. Phys. Lett A6 (1991) 635; J. Distler and P. Nelson, *New discrete states of strings near a black hole*, Nucl. Phys. **B374** (1992) 123; D. Gross and I. Klebanov, *S=1 for c=1*, Nucl. Phys. **B359** (1991) 3; J. Polchinski, *Ward identities in two-dimensional gravity*, Nucl. Phys. **B357** (1991) 241; E. Witten and B. Zwiebach, *Algebraic structures and differential geometry in 2d string theory*, Nucl. Phys. **B377** (1992) 55; hep-th/9201056.

11. L. Dolan and S. Horvath, *BRST properties of spin fields*, Nucl. Phys. **B448** (1995) 220, hep-th/9503210.

12. P. Goddard, private communication.

13. See also S.P. de Alwis, J. Polchinski and R. Schimmrigk, *Heterotic strings with tree level cosmological constant*, Phys. Lett. **B218** (1989) 449.

EXACT LOCAL SUPERSYMMETRY ABSENCE OF SUPERPARTNERS AND NONCOMMUTATIVE GEOMETRIES

Freydoon Mansouri*

Physics Department, University of Cincinnati
Cincinnati, OH 45221

Abstract

It is pointed out that if we allow for the possibility of a multilayered universe, it is possible to maintain exact supersymmetry and arrange, in principle, for the vanishing of the cosmological constant. Superpartners of a known particle will then be associated with other layers of such a universe. A concrete model realizing this scenario is exhibited in 2+1 dimensions, and it is suggested that it may be realizable in $3 + 1$ dimensions. The connection between this nonclassical geometry and noncommutative geometries is disscussed.

1   Introduction

Supersymmetry provides a rich and elegant theoretical framework for treating fermions and bosons on the same footing. It has been the basis of many developments for over two decades ranging from supersymmetric quantum mechanics [1] and supersymmetric gauge theories [2] to superstring theories [3]. The usefulness of this concept as an approximate symmetry in atomic [4] and nuclear physics [5] is already indisputable. What is not yet clear is whether it is a symmetry of Nature at the most fundamental level, and if so in what form.

From a purely theoretical point of view, the rich mathematical structure of super-symmetry has been used to address a number of important unsolved physical problems. Among these are the gauge hierarchy problem, the cosmological constant problem, and the dark matter problem. Moreover, making use of such concepts as duality, holomor-phicity, etc., supersymmetric gauge theories can be used to analyze the dynamics of gauge theories exactly [6]. This permits, among other things, a new approach to solving the longstanding strong coupling problem. Since the dynamical mechanisms made use of in these developments are standard to all gauge theories, it is hoped that they will also be applicable to non-supersymmetric gauge theories. It is thus clear that in the absence a competing framework general enough to address all of these problems, the

---

*e-mail address: mansouri@uc.edu

optimism in the relevance of some form of supersymmetry in a fundamental theory is not unreasonable.

One serious drawback of supersymmetry as a fundamental symmetry is its lack of experimental support. Up to the presently available energies, there is no evidence for the existence of the superpartners of the known fundamental particles such as the electron and the photon. The standard interpretation of the absence of superpartners is to assume that supersymmetry is spontaneously broken and that, as a result, the superpartners acquire large masses, making them undetectable at currently accessible energies. It then follows that the absence of superpartners is only temporary and that experiments at a high enough energy scale will eventually lead to their discovery. Depending on the particular model, a typical lower bound for such a scale is of the order of Tev's. Unfortunately, there are no reliable upper bounds for this scale below the Planck scale.

From the experience with flavor symmetry and the manner in which different generations of quarks and leptons were predicted and discovered at higher and higher energies, it is generally believed that if supersymmetry plays a fundamental role, the above interpretation is the logical next step beyond the bosonic symmetries in particle physics. On the other hand, in a broader perspective, the consequences of the manner in which supersymmetry is broken are not confined to the particle physics sector. They will also have profound cosmological consequences. In particular, if supersymmetry is spontaneously broken in the usual way, one would have to look for a different mechanism to ensure the vanishing of the cosmological constant. So, if we look to supersymmetry as basis for the simultaneous solution of both the cosmological constant problem and the absence of superpartners, we appear to have reached an impasse.

A way out of this dilemma was suggested by Witten [7] based on how local supersymmetry is realized in 2+1 dimensions. There is also an alternative unconventional suggestion by my collaborators and me [8], which is again based on how local supersymmetry is realized in 2+1 dimensions. In the following sections, I will describe, in turn, these suggestions, how the alternative view was arrived at, its connection to noncommutative geometry, and some of its consequences.

2 Witten's Observation

As mentioned in the in previous section, the jest of Witten's observation is that in 2+1 dimensions the requirement of local supersymmetry provides a mechanism that, at least in principle, ensures the vanishing of the cosmological constant without leading to equality of masses for the supersymmetric partners [7]. The success of this mechanism depends crucially on the manner in which the states of nonzero energy (mass) are realized in 2+1 dimensions. To see this, we note that states of nonzero energy produce geometries which are asymptotically conical [9]. To have Fermi- Bose degeneracy in such a conical geometry, supersymmetry must be realized linearly, i.e., we must have asymptotic supersymmetric generators (supercharges) connecting fermionic and bosonic states of a supermultiplet. On the other hand, supercharges transform as Lorentz spinors so that their existence depends on whether the corresponding manifold allows the construction of spinors which are asymptotically covariantly constant. This cannot happen in an asymptotically conical geometry [10]. As a result, there will be no supercharges for constructing a linear representation of supersymmetry to which fermionic and bosonic states of nonzero mass could belong. Therefore, there will be no Fermi-Bose mass degeneracy. On the other hand, the geometry produced by the vacuum state which is a state of zero mass is not asymptotically conical, there are no restrictions on spinors, the vacuum remains supersymmetric, and the cosmological can be made to vanish.

If it were possible to implement this mechanism in $3 + 1$ dimensions in a realistic manner, it wood significantly boost our near term confidence in the relevance of supersymmetry as a fundamental symmetry. The only obstacle on its way to full acceptance would then be its experimental confirmation.

3 An Alternative Proposal

In this section, I would like to present a point of view in which supersymmetry is realized in a supermultiplet of space-times which I will refer to as a supersymmetric spacetime. The geometry of such space-times are more complex than the familiar classical geometries and require the introduction of new concepts. It will be recalled that a classical metrical geometry is determined locally in terms of a differential line element or, equivalently, in terms of the components of a metric tensor. The supersymmetric space-time that we have in mind is an example of a nonclassical geometry which consists of the following elements : (i) The c-number line element is replaced with an "operator" line element. Equivalently, the components of the metric tensor are replaced with operators. (ii) These operators are constructed from the elements of an algebra. The particular cases of interest in the present context are Lie and super Lie algebras. (iii) There is an associated Hilbert space on which the elements of the algebra and the operators constructed from them act linearly. For a supersymmetric space-time, the corresponding Hilbert space is a supersymmetric multiplet realizing, say, the super Poincaré group.

The classical, long wave long wave length, limit of these nonclassical geometries can be determined by allowing the line element operator act on the states. Then the diagonal elements may be replaced by the corresponding eigenvalues. As a result, for each state of the Hilbert space, the line element operator produces a "layer" of classical space-time, the number, n, of layers being equal to the dimension of the (super) multiplet. The off-diagonal elements of the line element operator provide the means of communication among various layers. Thus in this limit, a nonclassical geometry consisting of n layers of d-dimensional classical geometries may be viewed as a $(d+1)$-dimensional geometry in which the range of one of the dimensions is finite and discrete. As an example, consider an $N = 2$ supersymmetric space-time. It consists of four layers of d-dimensional space-times in which different layers are related to each other by supersymmetry transformations. We will see below a concrete realization of this nonclassical geometry in 2+1 dimensions.

Although the nonclassical geometry described above appears to be general and independent of the dimension d, it is conceivable that, like the mechanism suggested by Witten [7], it will only have 2+1 dimensional realizations. But for the moment let us assume that it will also have $3 + 1$ dimensional realizations and consider some of its predictions. Representing a particle by a Poincaré state, we can put such a particle and its superpartners in a supermultiplet consisting of these Poincaré states. Then the above nonclassical geometry, in its simplest form, suggests that the particle and its superpartners reside in different layers of the supersymmetric space-time. Since, in the simplest model, supersymmetry is the only means of communication between the layers, to have any hope of obtaining information about the superpartners, supersymmetry must remain exact. From this it follows that a particle and its superpartner(s), if it can be called that, must have the same mass and that the cosmological constant problem is, in principle, solvable.

An immediate difficulty with this picture which comes to mind is that there is no experimental evidence for superpartners of the same mass. In this respect, we must note that the experiments in question were all perceived and carried out under the assumption of a single layered Universe. They are mostly "missing mass" type experiments. So one would not expect to obtain any information about the superpartners which "reside" in the other layers. Moreover, the very notion of a "superpartner" makes sense in a world of broken supersymmetry. In a superworld of exact supersymmetry, a particle and its superpartner(s) are different spin states of the same "superparticle". So one way of restating the lack of experimental evidence for the mass degenerate superpartners is to ask why it is that only one spin state of a superparticle appears in our experiments. A possible answer to this question is that a multilayered universe which emerges from a nonclassical geometry is very much like the many worlds picture necessary for an objective interpretation of quantum mechanics [11]. A superparticle in a multilayered universe is capable of being in any one of its spin states. In an experiment set up in any one layer, the wave function of the superparticle "collapses" into an eigenstate of spin consistent with that layer. This makes the task of obtaining information about superparticles highly non-trivial but not impossible. We must learn how quantum mechanics works in such a superworld. Needless to say, in the above discussion I have left out such intrinsic quantum mechanical effects as tunnelling, etc.

Finally, let me say a few words about a possible impact of a supersymmetric spacetime picture on the dark matter problem. In a one-layer universe to which we are accustomed, there is excellent experimental support for the equality of the gravitational and inertial masses. In a multilayered universe, there is no a' priori reason for equivalence principle to hold in its present form. So, it is plausible that the need for dark matter arises from the breakdown of the equivalence principle in a multilayered universe. In particular, a scheme in which the inertial masses are "rotated" with respect to gravitational masses may lead to the resolution of this problem.

4 Works on Noncommutative Geometries

The basic element of the nonclassical geometry described in the previous section is the introduction of an operator line element. The simplest way of viewing such an operator is to take the components of the metric tensor to be (noncommuting) operators. This statement is basis dependent, however, and a transformation to a different basis mixes the components of the metric tensor and the coordinates. Therefore, in the transformed basis the coordinates also become (noncommuting) operators. This means that we can view this nonclassical geometry as a form of a noncommutative geometry.

The subject of noncommutative geometry has appeared in theoretical physics in number of contexts. The most complete and consistent among these is the work of Connes [12]. From a purely physical point of view, it has appeared in the works of Witten [13] and of 't Hooft [14]. It is also inherent in any quantum mechanical matrix model, or zero-brain formalism such as the work of reference [15] and the references cited therein. To my knowledge, no systematic study has been undertaken to see whether or not all of these works as well as our nonclassical geometry fall within the general formalism of Connes. The answer to this question is likely to accelerate the progress in this field.

5 Lessons from 2+1 Dimensions

To provide a concrete realization of the nonclassical geometry discussed in the previous sections, we now turn to the Chern Simons gauge theory of the super Poincaré group in 2+1 dimensions. It has been known for sometime that supergravity theories in 2+1 dimensions can be formulated as Chern Simons gauge theories of the corresponding supergroups [16-19]. In this and the following sections, I will explore the physical properties of the emerging space-time when supersymmetric matter is coupled to these theories in a super Poincaré gauge invariant manner [20]. Let me begin with the simpler problem to the same aim, i.e., that of coupling matter to Poincaré Chern Simons gravity in a Poincaré gauge invariant manner. It has been known [20] for sometime that the two-body problem for this theory is exactly solvable. One of the important features of this approach is that the concept of space-time is not a fundamental input but an output of this gauge theory.

The general form of the Chern Simons action in 2+1 dimensions given by

$$I_{cs} = \int_M \gamma_{bc} A^b \wedge (dA^c + \frac{1}{3} f^c_{de} A^d \wedge A^e) \tag{1}$$

where A^a are components of the Lie algebra valued connection

$$A = A^a G_a; \quad A^a = A^a_\mu dx^\mu \tag{2}$$

The quantities G^a are elements of the Lie algebra with structure constants f_{abc}. The quantities γ_{ab} are the components of a suitable non-degenerate metric on the Lie algebra [17]. For Poincaré algebra with elements P_a, J_a, a=0,1,2, the connection can be written as

$$A_\mu = e^a_\mu P_a + \omega^a_\mu J_a; \quad \mu = 0, 1, 2 \tag{3}$$

where e^a_μ and ω^a_μ are gauge fields of the Poincaré group. The manifold M in Eq. 1 is not to be identified with the metrical space-time.

Consider next the coupling of the Chern Simons action to matter. Any coupling via matter fields appear to break the local Poincaré gauge symmetry to its Lorentz subgroup, so that we are limited to matter coupling via sources. The Poincaré invariance of the Chern Simons gauge theory suggests that we can introduce the notion of a particle or a source as an irreducible representation of the Poincaré group, in the same way as we do in particle physics in 3+1 dimensions. To this end, we take a particle to be an irreducible representation of the Poincaré group. Its first Casimir operator $p^2 = m^2$ determines the mass of the source, and its second Casimir operator $W^2 = m^2 s^2$ its spin s. So, for sources of any spin, the coupling to the Chern Simons action can be achieved in terms of the action [20]

$$I = \int_C d\tau \, \eta_{ab} \, [p^a \partial_\tau q^b + t^\mu (p^a e^b_\mu + j^a \omega^b_\mu)] + \lambda_1 (p^2 - m^2) + \lambda_2 (W^2 - m^2 s^2) \tag{4}$$

where $t^\mu = dx^\mu/d\tau$. It is clear from the action that the quantities p^a, and q^a are canonically conjugate to each other and satisfy Poisson brackets. For more than one source, we can add an action of this type for each one of them. In the presence of sources, the topology of the manifold M is modified, but the components of the field strength still vanish outside sources.

It has been shown that the problem of two sources coupled to the Chern Simons gravity can be solved by reducing it to an equivalent one-body problem [20]. This is done by taking full advantage of the topological features of the theory. In a topological gauge theory all the gauge invariant obsevables can be expressed in terms of

Wilson loops. By definition, the Wilson loop for the connection given by Eq. 2 in the representation $\mathcal{R}$ of the algebra is given by

$$W_{\mathcal{R}}(C) = Tr_{\mathcal{R}} P \exp\left[i \oint_C A\right] \tag{5}$$

where P denotes path ordering. The paths characterizing different Wilson loops are distinguished from each other not by their local coordinates but by their homotopy classes. Moreover, we note that the Casimir invariants of a Poincaré state, which we identify as mass and spin, must be Wilson loops. Thus we can view our gauge invariant observables of this theory as the Casimir invariants of an equivalent one-body Poincaré state. Such a source is source endowed with two charges: a charge $\Pi^a = (\Pi^0, \vec{\Pi})$ and a charge $\Psi^a = (\Psi^0, \vec{\Psi})$, such that the Casimir invariants of the corresponding state are given, respectively, by $\Pi \cdot \Pi = H^2$ and $\Pi \cdot \Psi = HS$. We identify H and S as the mass and spin of the one-body source and wish to evaluate them in terms of Wilson loops of the two body system. To do this, let $W_R(C_0)$ be a Wilson loop enclosing the equivalent one-body source. Since the dynamics generated by the reduced one-body formalism must be identical to that of the original two-body system, $W_R(C_0)$ must be equal to a Wilson loop, $W_R(C_{12})$, enclosing the two sources with charges (p_1^a, j_1^a) and (p_2^a, j_2^a), respectively. Thus, we require that [20]

$$W_R(C_0) = W_R(C_{12}) \tag{6}$$

The path C_{12} can be chosen uniquely by the requirement that the Wilson loop correctly represent the asymptotic observables of the emerging space-time.

The explicit evaluation of the Wilson loops were carried out in reference [20]. Here we quote the expression for H :

$$\cos\frac{H}{2} = \cos(\frac{m_1}{2})\cos(\frac{m_2}{2}) - \frac{p_1 \cdot p_2}{m_1 m_2}\sin(\frac{m_1}{2})\sin(\frac{m_2}{2}) \tag{7}$$

The Physical Space-Time

Let us now consider the structure of space-time which corresponds to this exact solution. Up to this point, we have constructed a Chern Simons gauge theory coupled to sources on $R \times \Sigma$ (x-space) which as we emphasized is metric independent and should not be identified with space-time. On the other hand, it is clear that the identification of quantities such as momenta and coordinates of physically realizable particles can only be made in a metrical space-time. So we must show how the notion of a metrical space-time emerges from this formalism and what our gauge invariant observables correspond to in such a space-time [21]. To this end, we recall that our two sources are characterized by charges (p_1^a, j_1^a) and (p_2^a, j_2^a) with the corresponding canonical coordinates q_1^a and q_2^a, respectively. Without loss of generality, let the first source be at rest at the origin, i. e. , $\vec{q}_1 = 0$. Then $\vec{q}_2 \equiv \vec{q}$ can be viewed as a relative coordinate. We parametrize $\vec{q}$ by its polar components: $\vec{q} = (r, \phi)$. By fixing $\vec{q}_1 = 0$, we have made a choice of gauge which fixes all the Poincaré gauge transformations except for the spatial rotations generated by J^0 and translations along q^0. To fix these, consider first the transformation

$$\vec{q}' = \left[\exp i\tau^0 J_0\right]\vec{q} \tag{8}$$

where

$$\tau^0 = (1 - \frac{H}{2\pi})\phi \equiv \alpha\phi = \phi' \tag{9}$$

Here H is the numerical value of the exact Hamiltonian given by Eq. 20. Being an element of Poincaré group, this transformation leaves the Casimir invariants H and S unchanged. But the resulting vector, $\vec{q'}$, is no longer 2π periodic and satisfies the matching conditions for the coordinates on a cone characterized by the deficit angle $\beta = H$. This can be seen by noting that the transformed coordinates $\vec{q'}$ acquire a phase under the rotation $\phi \rightarrow \phi + 2\pi$:

$$\vec{q'}(\phi + 2\pi) = [\exp{(2\pi - H)J_0}] \, \vec{q'}(\phi) \tag{10}$$

To completely fix the gauge, we must also fix translations along q^0. So, consider

$$q'^0 = q'^0(q^0, \phi') = q^0 - \frac{S\phi'}{2\pi\alpha} \tag{11}$$

It thus follows that the general reduction of the two-body problem to an equivalent one-body problem always leads, in a particular gauge, to the motion of the relative coordinate on a cone. We know from the analysis of metrical general relativity [9] that point sources generate conical space-times. For a single source, the deficit angle of the cone is determined by the energy (mass), E, of the source. We must therefore identify the quantity H with the total gravitational energy of the two body system. It generates a cone over which the relative coordinate of the reduced two body system moves. Despite their similarities, this cone should not be confused with the conical space of the test particle approximation. As is clear from Eq. 7, the deficit angle of our cone is determined by the Casimir invariant H which is a highly non-linear function of the masses and the momenta of the two sources. In terms of the gauge fixed variables, the expression for the line element takes the form

$$ds^2 = dq_0'^2 - dr^2 - r^2 d\phi'^2 \tag{12}$$

Or in terms of more familiar coordinates

$$ds^2 = (dq^0 - \frac{S d\phi}{2\pi})^2 - dr^2 - \alpha^2 r^2 d\phi^2 \tag{13}$$

The coordinates in these equivalent expressions are related by Eqs. 9 and 8. Aside from any specific significance associated with the quantities H and S in this context, (see below), Eqs. 12 and 13 are standard expressions for the line element of a spinning cone [9].

It is thus clear that it is not the manifold $R \times \Sigma$ (x- space) but the q-space, M_q, from which the classical space-time is manufactured. Once the spatial part of q^a is identified with the cone, relativistic invariance requires that q^0 be identified with the "classical time". The quantity H characterizing this space-time also supplies [20], as it should, the boundary term which is necessary for the consistency of the canonical formalism in the metrical theory [22]. Since, as we have noted, H depends non-trivially on the momenta of the two sources, then, because the components of the metric tensor given by Eq. 13 also depend explicitly on the canonically conjugate variables, i.e., q_1 and q_2, these components will have non-vanishing Poisson Brackets with each other. Moreover, it follows from Eqs. 9 and 11 that in the form given by Eq. 12 although the components of the metric are reduced to constants, the corresponding, primed, coordinates will have non-vanishing Poisson brackets. This suggests that, for consistency, the quantity S in Eq. 13, which is also a boundary term, should be replaced with $P.J/H$. This operators acts in the Hilbert space of the one-body Poincaré state. Thus we arrive at the "operator line element"

$$ds^2 = (dq^0 - \frac{P \cdot J d\phi}{2\pi H})^2 - dr^2 - \alpha^2 r^2 d\phi^2 \tag{14}$$

It is interesting to note that we can still write the line element operator in same form as that given by Eq. 12 if we define

$$q'^0 = q^0 - \frac{P \cdot J \phi}{2\pi H} \tag{15}$$

But then, as we have noted above, the coordinates in such a generalized geometry will no longer commute with each other.

In the classical large distance physics, the operators H and J may be safely replaced with their eigenvalues. But in a short distance quantum mechanical context, this geometrical non-commutativity cannot be ignored. As we will see in section 8, the operator interpretation of the line element will turn out to be crucial in describing the geometry of the supersymmetric space-time.

6 Supersources as Supersymmetry Multiplets

This section is devoted to the description of supersources which are to be coupled to the Chern Simons action for the super Poincaré group. It will be recalled [20] that in the case of Poincaré gravity the sources(particles) can be viewed as irreducible representations of the Poincaré group in the same way as these representations are used in particle physics in $3 + 1$ dimensions. Similarly, we take a superparticle(supersource) to be an irreducible representation of the super Poincaré group. From this point of view, a superparticle is an irreducible supermultiplet consisting of several Poincaré states related to each other by the action of the supersymmetry generators. Clearly, this can be done for a simple or an extended supersymmetry with or without central charges. But in the interest of explicitness, we will consider in detail the $N = 2$ super Poincaré group. The $N = 2$ super Poincaré algebra in 2+1 dimensions can be written as [23]

$$
\begin{array}{ll}
[J^a, J^b] = -i\epsilon^{abc} J_c & ; \quad [P^a, P^b] = 0 \\
[J^a, P^b] = -i\epsilon^{abc} P_c & ; \quad [P^a, Q_\alpha] = 0 \\
[J^a, Q_\alpha] = -(\sigma^a)_\alpha^{\ \beta} Q_\beta & ; \quad [P^a, Q'_\alpha] = 0 \\
[J^a, Q'_\alpha] = -(\sigma^a)_\alpha^{\ \beta} Q'_\beta & ; \quad \{Q_\alpha, Q_\beta\} = 0 \\
\{Q_\alpha, Q'_\beta\} = -\sigma^a_{\alpha\beta} P_a & ; \quad \{Q'_\alpha, Q'_\beta\} = 0 \\
a = 0, 1, 2 & ; \quad \alpha = 1, 2
\end{array}
\tag{16}
$$

The indices of the two component spinor charges Q_α and Q'_α are raised and lowered by the antisymmetric metric $\epsilon^{\alpha\beta}$ with $\epsilon^{12} = -\epsilon_{12} = 1$. the $SO(1,2)$ matrices σ^a satisfy the Clifford algebra

$$\{\sigma^a, \sigma^b\} = \frac{1}{2}\eta^{ab} \tag{17}$$

where η^{ab} is the Minkowski metric with signature $(+, -, -)$. We also have

$$\sigma^a_{\ \alpha\beta} = (\sigma^a)_\alpha^{\ \gamma} \epsilon_{\gamma\beta} \tag{18}$$

It is convenient to take the matrices σ^a to be

$$\sigma^0 = \frac{1}{2}\begin{pmatrix} 1 & 0 \\ 0 & -1 \end{pmatrix} \quad ; \quad \sigma^1 = \frac{1}{2}\begin{pmatrix} 0 & i \\ i & 0 \end{pmatrix} \quad ; \quad \sigma^2 = \frac{1}{2}\begin{pmatrix} 0 & 1 \\ -1 & 0 \end{pmatrix} \tag{19}$$

The two Casimir operators of the super Poincaré group are given by

$$C_1 = P^2 = \eta^{ab} P_a P_b \tag{20}$$

$$C_2 = \eta^{ab} P_a J_b + \epsilon^{\alpha\beta} Q'_\alpha Q_\beta \tag{21}$$

The first of these is the same as the Casimir operator of the Poincaré subgroup, so that its eigenvalues can be identified with the square of the mass of the superparticle. Since the Pauli-Lubanski operator (or its square) does not commute with supersymmetry transformations, it must be supplemented with the second term on the right hand side of Eq. 21 to obtain a super Poincaré invariant. We will designate its eigenvalues as mc_2.

Irreducible representations of the $N = 2$ super Poincaré group in 2+1 dimensions can be constructed along the same lines as those in 3+1 dimensions [24]. For massive states, without loss of generality we can work in a frame in which the supermultiplet is at rest. Then the non-vanishing anti-commutators of the superalgebra simplify to

$$\{Q_1, Q'_2\} = \{Q_2, Q'_1\} = \frac{m}{2} \tag{22}$$

Thus Q_α and Q'_α, $\alpha = 1, 2$, form a Clifford algebra. We define a Clifford vacuum state , $|\Omega>$ by the requirement

$$Q_\alpha|\Omega >= 0 \quad ; \quad \alpha = 1, 2 \tag{23}$$

It is easy to verify that such a state exists within every supermultiplet and that it is an eigenstate of C_1 and C_2:

$$C_1|\Omega > = m^2|\Omega > \tag{24}$$

$$C_2|\Omega > = mc_2|\Omega > \tag{25}$$

From the definition of the Clifford vacuum state in the rest frame of the superparticle, it follows that

$$\begin{aligned} C_2|\Omega > &= P \cdot J|\Omega > \\ &= ms^0|\Omega > \\ &= ms|\Omega > \end{aligned} \tag{26}$$

where we identify the eigenvalue, s, of the operator s^0 with the spin of the state $|\Omega >$. So, the Clifford vacuum state is a Poincaré state with mass m and spin s:

$$|\Omega >= |m, s > \tag{27}$$

Consider, next, the states

$$|\Omega_1 > = Q'_1|\Omega >, \tag{28}$$

$$|\Omega_2 > = Q'_2|\Omega > \tag{29}$$

and

$$|\Omega_{12} >= Q'_1 Q'_2|\Omega > \tag{30}$$

It is easy to verify that

$$s^0|\Omega_1 > = (s - \frac{1}{2})|\Omega_1 > \tag{31}$$

$$s^0|\Omega_2 > = (s + \frac{1}{2})|\Omega_1 > \tag{32}$$

$$s^0|\Omega_{12} > = s|\Omega_{12} > \tag{33}$$

These three Poincaré states together with the Clifford vacuum state form an Irreducible supermultiplet of $N = 2$ super Poincaré group in 2+1 dimensions, which we call a superparticle. Each supermultiplet is distinguished by its mass m and the eigenvalue $c_2 = s$, where s is the spin of the Clifford vacuum state. We will refer to c_2 as the superspin of the multiplet. The spins of the states within a supermultiplet are fixed once the value of c_2 is specified. For example, for $c_2 = \frac{1}{2}$, the resulting $N = 2$ supermultiplet is a vector multiplet consisting of a spin zero, two spin 1/2, and one spin one Poincaré states.

7   Exact Solution of the Two-Superbody Problem

It has been pointed out recently that the two-superbody problem in $N = 2$ Chern Simons supergravity is exactly solvable [23]. It was in the process of giving a physical interpretation to this solution that the departure from classical to nonclassical geometry became unavoidable [8]. I will briefly sketch the two superbody problem below and go over the supersymmetric space-time which emerges from it in the next section. As in Section 5, we begin with the general form of the Chern Simons action in 2+1 dimensions given by Eq. 1. In the present case, the quantities A^B are the components of the Lie superalgebra valued connection which for $N = 2$ super Poincaré algebra can be written as

$$A_\mu = e_\mu{}^a P_a + \omega_\mu{}^a J_a + \chi_\mu{}^\alpha Q_\alpha + \xi_\mu{}^\alpha Q'_\alpha \tag{34}$$

Then the covariant derivative is

$$D_\mu = \partial_\mu + iA_\mu \tag{35}$$

Then, just as in Poincaré gravity, the Chern Simons action for the super Poincaré group can be written as

$$
\begin{aligned}
I_{cs} = \ & \frac{1}{2} \int_M \{ \eta_{bc} [e^b \wedge (2d\omega^c + \epsilon^c{}_{da}\omega^d \wedge \omega^a)] \\
& -\epsilon_{\alpha\beta}[\chi^\alpha \wedge (d - i\sigma_a\omega^a)\psi^\beta + \psi^\alpha \wedge (d - i\sigma_a\omega^a)\chi^\beta]\}
\end{aligned}
\tag{36}
$$

As in the case of Poincaré gravity, the manifold M is specified by its topology and is not to be identified with space-time which will emerge (see below) as an output of this gauge theory.

To couple (super)sources to this Chern Simons theory, we proceed in a manner similar to the way sources were coupled to the Poincaré Chern Simons theory. From the discussion of the supermultiplets given in section 6, we conclude that the logical candidates for our supersources are the irreducible representations of the $N = 2$ super Poincaré group. Then each supersource can be coupled to the $N = 2$ Chern Simons supergravity by an action of the form [23]

$$
\begin{aligned}
I_s = \int_C d\tau \{ & p_a \partial_\tau q^a - \epsilon_{\alpha\beta} p^\alpha \partial_\tau q^\beta - t^\mu (e_\mu^a p_a + \omega_\mu^a j_a - i\epsilon_{\alpha\beta}\chi_\mu^\alpha p^\beta \\
& + (\sigma \cdot p)_{\alpha\beta} \xi_\mu^\alpha q^\beta) + \lambda_1 (p^2 - m^2) + \lambda_2 (c_2 - s) \}
\end{aligned}
\tag{37}
$$

where τ is an invariant parameter along the trajectory C. Also, mc_2 is an eigenvalue of the second Casimir operator of the super Poincaré group, and s is the spin of the Clifford vacuum state of the supermultiplet. The choice of the constraint multiplying λ_2 is crucial in relating the eigenvalue of the second Casimir invariant, c_2, of the superalgebra to the spin content of a supermultiplet. For more than one source, one can add an action of this type for each source. In the presence of supersources the topology of the manifold is modified. But the field strengths still vanish outside supersources, and the theory is locally trivial.

It was shown in reference [23] that the exact gauge invariant observables of the two-superbody system can be obtained in terms of Wilson loops. They may be viewed as the Casimir invariants of an equivalent one-superbody state, similar to the equivalent one-body state of Chern Simons gravity. We will refer to these invariants as H and C_2. As we have seen above, their eigenvalues determine mass(energy) and spin(angular momentum) content the supermultiplet. They constitute the asymptotic observables of the two-superbody system and were given in references [23]. Here we note that the expression for the invariant H is identical to the corresponding invariant for its Poincaré subgroup given by Eq. 7.

8   The Physical Space-Time

Having discussed the gauge invariant observables of the exact two-superbody system, we now turn to the structure of the corresponding space-time. We take our clue from the space-time structure which emerged in section 5 from the dynamics of the two-body system in Poincaré Chern Simons gravity. In the supersymmetric case, the situation is somewhat more complicated. To see why, we note that in both cases we can associate our gauge invariant observables to a reduced one-(super)body state. In the pure gravity case, such a state is a single Poincaré state. In the supersymmetric case it is a supermultiplet consisting of several (four for $N = 2$) Poincaré states. As we saw in section 5, in the case of Poincaré Chern Simons theory, the structure of the emerging space-time and its asymptotic observables are completely determined by the two (gauge invariant) Casimir invariants of the reduced one-body Poincaré state. To see how this picture generalizes for the two-superbody system, we recall that our two supersources are characterized by charges (p_1^A, j_1^A) and (p_2^A, j_2^A) with the corresponding canonical coordinates q_1^A and q_2^A, respectively. Without loss of generality, let the first supersource be at rest at the origin, i. e. , $\vec{q}_1 = 0$. Then $\vec{q}_2 \equiv \vec{q}$ can be viewed as a relative coordinate. As in pure gravity, we parametrize $\vec{q}$ by its polar components: $\vec{q} = (r, \phi)$. By fixing $\vec{q}_1 = 0$, we have again made a choice of gauge which fixes all the $N = 2$ super Poincaré gauge transformations except for the rotations generated by J^0 and translations along q^0. To fix these, consider first the same transformation as that specified by Eqs. 8 and 9. Being an element of $N = 2$ super Poincaré group, this transformation leaves the Casimir invariants H and C_2 unchanged. But again the $\vec{q}'$ is no longer 2π periodic and satisfies the matching conditions for the coordinates on a cone characterized by the deficit angle $\beta = H$.

Up to this point, everything looks the same as in Poincaré gravity discussed in Section 5. However, differences begin to appear when we try to gauge fix the translations along q^0. It will be recalled from our discussion of supersources that an $N = 2$ supermultiplet at rest with Casimir invariants H and C_2 consists of four Poincaré states. Writing the eigenvalues of C_2 as Hc for the Clifford vacuum, these four states will have the following spin eigenvalues :

$$P \cdot J|H, c, s_1 > \ = \ H(c - \frac{1}{2})|H, c_2, s_1 > \tag{38}$$

$$P \cdot J|H, c, s_2 > \ = \ Hc|H, c, s_2 > \tag{39}$$

$$P \cdot J|H, c, s_3 > \ = \ Hc|H, c, s_3 > \tag{40}$$

$$P \cdot J|H, c, s_4 > \ = \ H(c + \frac{1}{2})|H, c, s_4 > \tag{41}$$

In the case of Poincaré Chern Simons gravity, it was possible to also fix the gauge in q^0 direction by the transformation given by Eq. 11 which involved the spin of the Poincaré state. Clearly, this is no longer possible for a supermultiplet consisting of Poincaré states of different spin. This makes it impossible for a single c-number line element of the form given by Eqs. 12 and 13 to describe all the spin states of our equivalent one-superbody multiplet even in the case of classical large distance gravity. So, to describe all the spin states corresponding to our gauge invariant observables H and C_2, we must generalize the usual notion of a c-number line element to the "operator line element" given by Eq. 14, which now acts on the Poincaré states making up the supermultiplet. When this operator line element acts on a state of a supermultiplet, we can replace, at least for large distance physics, the operator $P.J/H$ with the spin eigenvalue of that state and hence specify the corresponding c-number space-time. It therefore follows that the description of all the spin states of the equivalent one-body supermultiplet requires a multiplet of space-times equal in number to the dimension of the supermultiplet (four for $N = 2$). With $k = 1, .., 4$, the line elements for the members of this space-time multiplet are given by

$$ds_k^2 = (dq^0 - \frac{s_k d\phi}{2\pi})^2 - dr^2 - \alpha^2 r^2 d\phi^2 \tag{42}$$

The line element operator in Eq. 14 is not invariant under supersymmetry transformations, and it transforms in the same way as the Poincaré states within a supermultiplet. In other words, for $k = 1, .., 4$ the line elements in Eq. 42 form an irreducible representation of the $N = 2$ supersymmetry and are completely determined by the asymptotic observables H and C_2. Thus, the metrical description of the two-super-body system coupled to the super Poincaré Chern Simons action requires not just one but a supermultiplet of space-times. The supersymmetry generators act as ladder operators relating different layers of this nonclassical geometry. In its simplest form such as in the classical large distance regime, this supersymmetric space-time may be viewed as an ordinary space-time with an additional finite discrete dimension.

We have thus verified that the supersymmetric space-time which emerges from the exact solution of the two-superbody in 2+1 dimensions is, in all details, a realization of the nonclassical geometry discussed in section 3.

9   Concluding Remarks

Like Witten's suggestion described in section 2, it might be thought that the interesting applications of the nonclassical geometry described in this work are confined to 2+1 dimensions. This may well turn out to be the case. However, it is not difficult to conceive of $3 + 1$ dimensional realizations of this geometry, which may or may not be interesting. For example, noting the correspondence between point-like sources in 2+1 dimensions and infinite line sources in $3 + 1$ dimensions, one can extend the supersymmetric space-time discussed in the previous section to $3 + 1$ dimensions by simply adding a dz^2 term to each line element in Eq. 42. Work is in progress to see how one can detect experimentally the multilayer effects of a supersymmetric space-time. For one thing, this may also a way of testing the many worlds picture of quantum mechanics. Clearly, much remains to be clarified.

ACKNOWLEDGMENT

This work was supported in part by the Department of Energy under the contract No. DOE-FG02-84ER40153.

References

1. E. Witten, *Nucl. Phys.* **B188** (1981) 513.

2. J. Wess and B. Zumino, *Nucl. Phys.* **B70** (1974) 39.

3. M.B. Green, J.H. Schwarz, E. Witten, *Superstrings*, **Vol. I, Vol. II**, Cambridge University Press, 1985.

4. V.A. Kostelecky and M.M. Nieto, *Phys. Rev.* **A32** (1985) 1293.

5. F. Iyacello, *Phys. Rev. Lett.* **44** (1980) 772.

6. N. Seiberg and E. Witten, *Nucl. Phys.* **B426** (1994) 19, **431** (1994) 484.

7. E. Witten, *Int. Jour. Mod. Phys.* **A10** (1995) 1247.

8. A preliminary version of our results were first reported at *The International Conference on Seventy Years of Quantum Mechanics and Modern Trends in Theoretical Physics*, Calcutta, India 1/29- 2/2/1996, to appear in the Proceedings, ed. P. Bandyopadhyay; Sunme Kim and Freydoon Mansouri, e-print gr-qc/9609037, Physics Letters, *in press*; F. Ardalan, S. Kim, F. Mansouri, *Int. Jour. Mod. Phys.* **A12** (1997) 1183

9. S.Deser, R.Jackiw, G. 't Hooft, *Ann. of Phys.* (N.Y.) **152** (1984) 220; S. Giddings, J. Abbott and K. Kuchar, *Gen. Rel. Grav.* **16**(1984) 751; J.R. Gott and M. Alpert, *Gen. Rel. Grav.* **16** (1984) 751.

10. M. Henneaux, *Phys. Rev.* **D29** (1984) 2766

11. H. Everett, III, in *The Many Worlds Interpretation of Quantum Mechanics*, ed. by B.S. DeWitt and N. Graham, Princeton University Press, 1973.

12. A. Connes, *Non-commutative Geometry*, Academic Press, 1994

13. E. Witten, *Nucl. Phys.* **B460** (1995) 335.

14. G. 't Hooft, Utrecht preprint THU-96/02, e-print gr-qc/9601014

15. T. Banks, W. Fischler, S.H. Shenker, L. Susskind, e-print hep-th/9610043.

16. A. Achucarro and P.K. Townsend, *Phys. Lett* **B180** (1986) 89

17. E. Witten, *Nucl. Phys.* **B311** (1988) 46 and **B323** (1989) 113

18. K. Koehler, F. Mansouri, C. Vaz, L. Witten, *Mod. Phys. Lett.* **A5**(1990) 935

19. K. Koehler, F. Mansouri, C. Vaz, L. Witten, *Nucl. Phys.* **B341** (1990) 167 and**B348** (1990) 373

20. F. Mansouri and M.K. Falbo-Kenkel, *Mod. Phys. Lett.*; *Jour. Math. Phys.* **A8** (1993) 2503; F. Mansouri, *Comm. Theo. Phys.* **4** (1995) 191.

21. F. Mansouri in *Proceedings of Twenty Third Coral Gables Conference*, ed. B. Kursunoglu, S. Mintz, A. Perlmutter, Plenum Press, 1995

22. T. Regge and C. Teitelboim, *Ann. Phys.* **88** (1974) 236

23. Sunme Kim and F. Mansouri, *Phys. Lett.* **B 372**(1996) 72

24. A. Salam and J. Strathdee, *Forts. Phys.* 26 (1978) 57; P.G.O. Freund, *Introduction to Supersymmetry*, Cambridge University Press, 1986; J. Wess and J. Bagger, *Supersymmetry and Supergravity*, second edition, Princeton University press, 1992

SECTION VI
LIGHT CONE QUANTIZATION

ADJOINT QCD$_2$ IN LARGE N

Stephen Pinsky

Department of Physics
The Ohio State University
174 West 18th Avenue
Columbus, Ohio 43210

Abstract

We consider a dimensional reduction of 3+1 dimensional $SU(N)$ Yang-Mills theory coupled to adjoint fermions to obtain a class of $1+1$ dimensional gauge theories. We derive the quantized light-cone Hamiltonian in the light-cone gauge $A_- = 0$ and large-N limit, then solve for the masses, wavefunctions and of the color singlet boson and fermion boundstates. We find that the theory has many exact massless state that are similar to the t'Hooft pion.

INTRODUCTION

In this work [1], we start by considering QCD$_{3+1}$ coupled to Dirac adjoint fermions. Here, the virtual creation of fermion-antifermion pairs is not suppressed in the large-N limit – in contrast to the case for fermions in the fundamental representation [2] – and so one may study the structure of boundstates beyond the valence quark (or quenched) approximation[1]. We also anticipate that the techniques employed here will have special interest in the context of solving supersymmetric matrix theories.

The QCD$_{3+1}$ theory coupled to adjoint fermions is reduced to a $1+1$ dimensional field theory by stipulating that all fields are independent of the transverse coordinates $x^\perp = (x^1, x^2)$. The resulting theory is QCD$_{1+1}$ coupled to two $1+1$ dimensional complex adjoint spinor fields, and two real adjoint scalars. A key strategy in formulating this model field theories is to retain as many of the essential degrees of freedom of higher dimensional QCD while still being able to extract complete non-perturbative solutions. One finds Yukawa interactions between the scalars and fermion fields. While this approach is not equivalent to solving the full $3+1$ theory and then going to the regime where $k_\perp$ is relatively small, it may share many qualitative features of the higher dimensional theory, since the longitudinal dynamics is treated exactly. Studies of this type for pure glue and with fundamental quarks have yielded a number of interesting results [2, 3, 4].

The unique features of light-front quantization [5] make it a powerful tool for the nonperturbative study of quantum field theories. The main advantage of this approach is the apparent simplicity of the vacuum state. Indeed, naive kinematic arguments suggest that the physical vacuum is trivial on the light front. Since in this case all fields transform in the adjoint representation of $SU(N)$, the gauge group of the theory is actually $SU(N)/Z_N$, which has nontrivial topology and vacuum structure. For the particular gauge group $SU(2)$ this has been discussed elsewhere [6]. While this vacuum structure may in fact be relevant for a discussion on condensates, for the purposes of this calculation they will be ignored.

In the first section we formulate the $3+1$ dimensional $SU(N)$ Yang-Mills theory and then perform dimensional reduction to obtain a $1+1$ dimensional matrix field theory. The light-cone Hamiltonian is then derived for the light-cone gauge $A_- = 0$ following a discussion of the physical degrees of freedom of the theory. Singularities from Coulomb interactions are regularized in a natural way, and we outline how particular "ladder-relations" take care of potentially troubling singularities for vanishing longitudinal momenta $k^+ = 0$. In the final section exact massless solutions of the boundstate integral equations are discussed.

DEFINITIONS

We first consider $3+1$ dimensional $SU(N)$ Yang-Mills coupled to a Dirac spinor field whose components transform in the adjoint representation of $SU(N)$:

$$\mathcal{L} = \text{Tr}\left[-\frac{1}{4} F_{\mu\nu} F^{\mu\nu} + \frac{i}{2}(\bar{\Psi}\gamma^\mu \overleftrightarrow{D}_\mu \Psi) - m\bar{\Psi}\Psi \right] , \tag{1}$$

where $D_\mu = \partial_\mu + ig[A_\mu, \ ]$ and $F_{\mu\nu} = \partial_\mu A_\nu - \partial_\nu A_\mu + ig[A_\mu, A_\nu]$. We also write $A_\mu = A_\mu^a \tau^a$ where τ^a is normalized such that $\text{Tr}(\tau^a \tau^b) = \delta_{ab}$. The projection operators[1] Λ_L, Λ_R permit a decomposition of the spinor field $\Psi = \Psi_L + \Psi_R$,

---

[1] We use the conventions $\gamma^\pm = (\gamma^0 \pm \gamma^3)/\sqrt{2}$, and $x^\pm = (x^0 \pm x^3)/\sqrt{2}$.

where

$$\Lambda_L = \frac{1}{2}\gamma^+\gamma^-, \quad \Lambda_R = \frac{1}{2}\gamma^-\gamma^+ \quad \text{and} \quad \Psi_L = \Lambda_L\Psi, \quad \Psi_R = \Lambda_R\Psi. \quad (2)$$

Inverting the equation of motion for Ψ_L, we find

$$\Psi_L = \frac{1}{2iD_-}\left[i\gamma^i D_i + m\right]\gamma^+\Psi_R \quad (3)$$

where $i = 1, 2$ runs over transverse space. Therefore Ψ_L is not an independent degree of freedom.

Dimensional reduction of the $3 + 1$ dimensional Lagrangian (1) is performed by assuming (at the classical level) that all fields are independent of the transverse coordinates $x^\perp = (x^1, x^2)$: $\partial_\perp A_\mu = 0$ and $\partial_\perp \Psi = 0$. In the resulting $1 + 1$ dimensional field theory, the transverse components $A_\perp = (A_1, A_2)$ of the gluon field will be represented by the $N \times N$ complex matrix fields $\phi_\pm$:

$$\phi_\pm = \frac{A_1 \mp iA_2}{\sqrt{2}}. \quad (4)$$

Here, ϕ_- is just the Hermitian conjugate of ϕ_+. When the theory is quantized, $\phi_\pm$ will correspond to ± 1 helicity bosons (respectively).

The components of the Dirac spinor Ψ are the $N \times N$ *complex* matrices $u_\pm$ and $v_\pm$, which are related to the left and right-moving spinor fields according to

$$\Psi_R = \frac{1}{2^{\frac{1}{4}}}\begin{pmatrix} u_+ \\ 0 \\ 0 \\ u_- \end{pmatrix} \quad \Psi_L = \frac{1}{2^{\frac{1}{4}}}\begin{pmatrix} 0 \\ v_+ \\ v_- \\ 0 \end{pmatrix} \quad (5)$$

Adopting the light-cone gauge $A_- = 0$ allows one to explicitly rewrite the left-moving fermion fields $v_\pm$ in terms of the right-moving fields $u_\pm$ and boson fields $\phi_\pm$, by virtue of equation (3). We may therefore eliminate $v_\pm$ dependence from the field theory. Moreover, Gauss' Law

$$\partial_-^2 A_+ = g\left(i[\phi_+, \partial_-\phi_-] + i[\phi_-, \partial_-\phi_+] + \{u_+, u_+^\dagger\} + \{u_-, u_-^\dagger\}\right) \quad (6)$$

permits one to remove any explicit dependence on A_+, and so the remaining *physical* degrees of freedom of the field theory are represented by the helicity $\pm\frac{1}{2}$ fermions $u_\pm$, and the helicity ± 1 bosons $\phi_\pm$. There are no ghosts in the quantization scheme adopted here. In the light-cone frame the Poincaré generators P^- and P^+ for the reduced $1 + 1$ dimensional field theory are given by

$$P^+ = \int_{-\infty}^{\infty} dx^- \text{Tr}\left[2\partial_-\phi_- \cdot \partial_-\phi_+ + \frac{i}{2}\sum_h \left(u_h^\dagger \cdot \partial_-u_h - \partial_-u_h^\dagger \cdot u_h\right)\right] \quad (7)$$

$$P^- = \int_{-\infty}^{\infty} dx^- \text{Tr} \left[m_b^2 \phi_+ \phi_- - \frac{g^2}{2} J^+ \frac{1}{\partial_-^2} J^+ + \frac{tg^2}{2} [\phi_+, \phi_-]^2 + \sum_h F_h^\dagger \frac{1}{i\partial_-} F_h^\dagger \right] \quad (8)$$

where the sum $\sum_h$ is over $h = \pm$ helicity labels, and

$$J^+ = \text{i}[\phi_+, \partial_- \phi_-] + \text{i}[\phi_-, \partial_- \phi_+] + \{u_+, u_+^\dagger\} + \{u_-, u_-^\dagger\} \quad (9)$$

$$F_\pm = \mp sg [\phi_\pm, u_\mp] + \frac{m}{\sqrt{2}} u_\pm \quad (10)$$

We have generalized the couplings by introducing the variables t and s, which do not spoil the $1 + 1$ dimensional gauge invariance of the reduced theory; the variable t will determine the strength of the quartic-like interactions, and the variable s will determine the strength of the Yukawa interactions between the fermion and boson fields, and appears explicitly in equation (10). The dimensional reduction of the original $3 + 1$ dimensional theory yields the canonical values $s = t = 1$.

Renormalizability of the reduced theory also requires the addition of a bare coupling m_b, which leaves the $1+1$ dimensional gauge invariance intact. In all calculations, the renormalized boson mass $\tilde{m}_b$ will be set to zero.

Canonical quantization of the field theory is performed by decomposing the boson and fermion fields into Fourier expansions at fixed light-cone time $x^+ = 0$:

$$u_\pm = \frac{1}{\sqrt{2\pi}} \int_{-\infty}^{\infty} dk \, b_\pm(k) e^{-\text{i}kx^-} \quad \text{and} \quad \phi_\pm = \frac{1}{\sqrt{2\pi}} \int_{-\infty}^{\infty} \frac{dk}{\sqrt{2|k|}} a_\pm(k) e^{-\text{i}kx^-}$$
$$\quad (11)$$

where $b_\pm = b_\pm^a \tau^a$ etc. We also define

$$b_\pm(-k) = d_\mp^\dagger(k), \quad a_\pm(-k) = a_\mp^\dagger(k), \quad (12)$$

where $d_\pm$ correspond to antifermions. Note that $(b_\pm^\dagger)_{ij}$ should be distinguished from $b_{\pm ij}^\dagger$, since in the former the quantum conjugate operator $\dagger$ acts on (color) indices, while it does not in the latter. The latter formalism is sometimes customary in the study of matrix models. The precise connection between the usual gauge theory and matrix theory formalism may be stated as follows:

$$b_{\pm ji}^\dagger = b_\pm^{a\dagger} \tau_{ji}^{a*} = b_\pm^{a\dagger} \tau_{ij}^a = (b_\pm^\dagger)_{ij}$$

The commutation and anti-commutation relations (in matrix formalism) for the boson and fermion fields take the following form in the large-N limit $(k, \tilde{k} > 0; h, h' = \pm)$:

$$\left[a_{hij}(k), a_{h'kl}^\dagger(\tilde{k}) \right] = \{ b_{hij}(k), b_{h'kl}^\dagger(\tilde{k}) \} = \{ d_{hij}(k), d_{h'kl}^\dagger(\tilde{k}) \} = \delta_{hh'} \delta_{jl} \delta_{ik} \delta(k - \tilde{k}),$$
$$\quad (13)$$

where have used the relation $\tau_{ij}^a \tau_{kl}^a = \delta_{il} \delta_{jk} - \frac{1}{N} \delta_{ij} \delta_{kl}$. All other (anti)commutators vanish.

The Fock space of physical states is generated by the color singlet states, which have a natural 'closed-string' interpretation. They are formed by a color trace of the fermion, antifermion and boson operators acting on the vacuum state $|0\rangle$. Multiple string states couple to the theory with strength $1/N$, and so may be ignored.

THE LIGHT CONE HAMILTONIAN

For the special case $\tilde{m}_b = m = t = s = 0$, the light-cone Hamiltonian is simply given by the current-current term $J^+\frac{1}{\partial^2_-}J^+$ in equation (8). In momentum space, this Hamiltonian takes the form

$$
\begin{aligned}
P^-_{J^+.J^+} &= \frac{g^2}{2\pi} \int_{-\infty}^{\infty} dk_1 dk_2 dk_3 dk_4 \frac{\delta(k_1 + k_2 - k_3 - k_4)}{(k_3 - k_1)^2} \frac{\text{Tr}}{2} \Bigg[\\
&\quad \sum_{h,h'} : \{b^\dagger_h(k_1), b_h(k_3)\} :: \{b^\dagger_{h'}(k_2), b_{h'}(k_4)\} : \\
&\quad + \frac{(k_1 + k_3)(k_2 + k_4)}{4\sqrt{|k_1||k_2||k_3||k_4|}} : [a^\dagger_+(k_1), a_+(k_3)] :: [a^\dagger_+(k_2), a_+(k_4)] : \\
&\quad + \frac{(k_2 + k_4)}{2\sqrt{|k_2||k_4|}} \sum_h : \{b^\dagger_h(k_1), b_h(k_3)\} :: [a^\dagger_+(k_2), a_+(k_4)] : \\
&\quad + \frac{(k_3 + k_1)}{2\sqrt{|k_1||k_3|}} \sum_{h'} : [a^\dagger_+(k_1), a_+(k_3)] :: \{b^\dagger_{h'}(k_2), b_{h'}(k_4)\} : \Bigg] \quad (14)
\end{aligned}
$$

The explicit form of the Hamiltonian (14) in terms of the operators $b_\pm$, $d_\pm$ and $a_\pm$ is straightforward to calculate, but too long to be written down here. It should be stressed, however, that several $2 \to 2$ parton processes are suppressed by a factor $1/N$, and so are ignored in the large-N limit. No terms involving $1 \leftrightarrow 3$ parton interactions are suppressed in this limit, however.

One can show that this Hamiltonian conserves total helicity h, which is an additive quantum number. Moreover, the number of fermions *minus* the number of antifermions is also conserved in each interaction, and so we have an additional quantum number $\mathcal{N}$. States with $\mathcal{N} = even$ will be referred to as *boson* boundstates, while the quantum number $\mathcal{N} = odd$ will refer to *fermion* boundstates. We will pay special attention to the cases $\mathcal{N} = 0$ and 3, since the associated states appear to be analogous to conventional mesons and baryons (respectively).

The instantaneous Coulomb interactions involving $2 \to 2$ parton interactions behave singularly when there is a zero exchange of momentum between identical 'in' and 'out' states. The same type of Coulomb singularity involving $2 \to 2$ boson-boson interactions appeared in a much simpler model [7], and can be shown to cancel a 'self-induced' mass term (or self-energy) obtained from normal ordering the Hamiltonian. The same prescription works in the model studied here. There are also finite residual terms left over after this cancellation is explicitly performed for the boson-boson and boson-fermion interactions, and they cannot be absorbed by a redefinition of existing coupling constants. These residual terms behave as momentum-dependent mass terms, and in some sense represent the flux-tube energy between adjacent partons in a color singlet state. For the boson-boson and boson-fermion interactions they are respectively

$$\frac{g^2 N}{2\pi} \cdot \frac{\pi}{4\sqrt{k_b k_{b'}}} \quad \text{and} \quad \frac{g^2 N}{2\pi} \frac{1}{k_f} \left(\sqrt{1 + \frac{k_f}{k_b}} - 1 \right) \tag{15}$$

where k_b, k_b' denote boson momenta, and k_f denotes a fermion momentum. These terms simply multiply the wavefunctions in the boundstate integral equations.

If we now include the contributions $F_h^\dagger \frac{1}{i\partial_-} F_h$ in the light-cone Hamiltonian (8), then we will encounter another type of singularity for vanishing longitudinal momenta $k^+ = 0$. This singular behavior can be shown to cancel a (divergent) momentum-dependent mass term, which is obtained after normal ordering the $F_h^\dagger \frac{1}{i\partial_-} F_h$ interactions and performing an appropriate (infinite) renormalisation of the bare coupling m_b. This momentum-dependent mass term has the explicit form

$$\frac{s^2 g^2 N}{2\pi} \int_0^\infty dk_1 dk_2 \left\{ \left(\frac{1}{k_2(k_1 - k_2)} + \frac{1}{k_2(k_1 + k_2)} \right) \sum_h a_h^\dagger(k_1) a_h(k_1) \right.$$
$$\left. + \frac{1}{k_2(k_1 - k_2)} \sum_h b_h^\dagger(k_1) b_h(k_1) + \frac{1}{k_2(k_1 + k_2)} \sum_h d_h^\dagger(k_1) d_h(k_1) \right\} \tag{16}$$

The mechanism for cancellation here is different from the Coulombic case, since we will require specific endpoint relations relating different wavefunctions. Before outlining the general prescription for implementing this cancellation, we consider a simple rendering of the boundstate integral equations involving the $F_h^\dagger \frac{1}{i\partial_-} F_h$ interactions. In particular, let us consider the helicity zero sector with $\mathcal{N} = 0$, and allow at most three partons. Then the boundstate integral equation governing the behavior of the wavefunction $f_{a_+ a_-}(k_1, k_2)$ for the two-boson state $\frac{1}{N} \text{Tr}[a_+^\dagger(k_1) a_-^\dagger(k_2)]|0\rangle$ takes the form

$$M^2 f_{a_+ a_-}(x_1, x_2) = \frac{g^2 N}{\pi} \cdot \frac{\pi}{4\sqrt{x_1 x_2}} f_{a_+ a_-}(x_1, x_2)$$
$$+ \frac{s^2 g^2 N}{\pi} \sum_{i=1,2} \int_0^\infty dy \left(\frac{1}{y(x_i - y)} + \frac{1}{y(x_i + y)} \right) f_{a_+ a_-}(x_1, x_2)$$
$$- msg \sqrt{\frac{N}{2\pi}} \int_0^\infty d\alpha d\beta \, \delta(\alpha + \beta - x_1) \times$$
$$\frac{1}{\sqrt{x_1}} \left(\frac{1}{\alpha} + \frac{1}{\beta} \right) \left[f_{b_+ d_+ a_-}(\alpha, \beta, x_2) + f_{d_+ b_+ a_-}(\alpha, \beta, x_2) \right]$$
$$+ \dots \tag{17}$$

where $M^2 = 2P^+P^-$, and $x_i = k_i/P^+$ are (boost invariant) longitudinal momentum fractions. Evidently, the integral (17) arising from $1 \to 2$ parton interactions behaves singularly for vanishing longitudinal momentum fraction $\alpha \to 0$, or $\beta \to 0$. However, these divergences are precisely canceled by the momentum-dependent mass terms, which represent the contribution (16).

To see this, we may consider the integral equation governing the wavefunction $f_{b_+d_+a_-}(k_1, k_2, k_3)$ for the three-parton state $\frac{1}{N^{3/2}}\text{Tr}[b_+^\dagger(k_1)d_+^\dagger(k_2)a_-^\dagger(k_3)]|0\rangle$:

$$
\begin{aligned}
M^2 f_{b_+d_+a_-}(x_1, x_2, x_3) &= m^2 \left(\frac{1}{x_1} + \frac{1}{x_2}\right) f_{b_+d_+a_-}(x_1, x_2, x_3) \\
&+ \frac{g^2 N}{\pi} \sum_{i=1,2} \left[\frac{1}{x_i}\left(\sqrt{1 + \frac{x_i}{x_3}} - 1\right)\right] f_{b_+d_+a_-}(x_1, x_2, x_3) \\
&- msg\sqrt{\frac{N}{2\pi}} \frac{1}{\sqrt{x_1 + x_2}} \left(\frac{1}{x_1} + \frac{1}{x_2}\right) f_{a_+a_-}(x_1 + x_2, x_3) \\
&+ \ldots
\end{aligned}
\tag{18}
$$

If we now multiply both sides of the above equation by x_i, and then let $x_i \to 0$ for $i = 1, 2$, we deduce the relations

$$
f_{b_+d_+a_-}(0, x_2, x_3) = \frac{sg}{m}\sqrt{\frac{N}{2\pi}} \frac{f_{a_+a_-}(x_2, x_3)}{\sqrt{x_2}}
\tag{19}
$$

$$
f_{b_+d_+a_-}(x_1, 0, x_3) = \frac{sg}{m}\sqrt{\frac{N}{2\pi}} \frac{f_{a_+a_-}(x_1, x_3)}{\sqrt{x_1}}
\tag{20}
$$

It is now straightforward to show that the singular behavior of the integral (17) involving the wavefunction $f_{b_+d_+a_-}$ may be written in terms of a momentum-dependent mass term involving the wavefunction $f_{a_+a_-}$. Similar divergent contributions are obtained from the the wavefunctions $f_{d_+b_+a_-}$, $f_{a_+b_-d_-}$ and $f_{a_+d_-b_-}$, all of which may be re-expressed in terms of the wavefunction $f_{a_+a_-}$ by virtue of corresponding 'ladder relations'. The sum of these divergent contributions exactly cancels the self-energy contribution. An entirely analogous set of ladder relations were found for the case of fermions in the fundamental representation of $SU(N)$ [2].

For the general case where states are permitted to have more than three partons, the correct ladder relations are not immediately obvious from an analysis of the integral equations alone. Nevertheless, they may be readily obtained from the constraint equation governing the left-moving fermion field Ψ_L. In particular, we have $i\partial_- v_\mp = F_\pm$, and so vanishing fields at spatial infinity would imply

$$
\int_{-\infty}^{\infty} dx^- F_\pm |\Psi\rangle = 0
\tag{21}
$$

for color singlet states $|\Psi\rangle$. The analysis of this condition in momentum space is quite delicate, since it involves integrals of singular wavefunctions over spaces of measure zero [8]. Viewed in this way we see that the ladder relations are the continuum equivalent of zero mode constraint equations that have shown to lead to spontaneous symmetry breaking in discrete light-cone quantization [9].

EXACT SOLUTIONS

For the special case $s = m = \tilde{m}_b = 0$, the only surviving terms in the Hamiltonian (8) are the current-current interactions $J^+ \frac{1}{\partial^2} J^+$ and the ϕ^4 interaction. This theory has infinitely many massless boundstates, and the partons in these states are either fermions or antifermions. States with bosonic $a_\pm$ quanta are always massive. One also finds that the massless states are pure, in the sense that the number of partons is a fixed integer, and there is no mixing between sectors of different parton number. In particular, for each integer $n \geq 2$, one can always find a massless boundstate consisting of a superposition of only n-parton states. A striking feature is that the wavefunctions of these states are *constant*, and so these states are natural generalizations of the constant wavefunction solution appearing in t'Hooft's model [10].

We present an explicit example below of such a constant wavefunction solution involving a three fermion state with total helicity $+\frac{3}{2}$, which is perhaps the simplest case to study. Massless states with five or more partons appear to have more than one wavefunction which are non-zero and constant, and in general the wavefunctions are unequal. It would be interesting to classify all states systematically, and we leave this to future work. One can, however, easily count the number of massless states. In particular, for $\mathcal{N} = 3$, $h = +\frac{3}{2}$ states, there is one three-parton state, 2 five-parton states, 14 seven-parton states and 106 nine-parton states that yield massless solutions.

Let us now consider the action of the light-cone Hamiltonian P^- on the three-parton state

$$
|b_+ b_+ b_+ \rangle = \int_0^\infty dk_1 dk_2 dk_3 \, \delta(\sum_{i=1}^3 k_i - P^+) f_{b_+ b_+ b_+}(k_1, k_2, k_3)
$$
$$
\frac{1}{N^{3/2}} \text{Tr}[b_+^\dagger(k_1) b_+^\dagger(k_2) b_+^\dagger(k_3)]|0\rangle \tag{22}
$$

The quantum number $\mathcal{N}$ is 3 in this case, and ensures that the state $P^- |b_+ b_+ b_+\rangle$ must have at least three partons. In fact, one can deduce the following:

$$
P^- | b_+ b_+ b_+ \rangle = \int_0^\infty dk_1 dk_2 dk_3 \, \delta(\sum_{i=1}^3 k_i - P^+)
$$
$$
-\frac{g^2 N}{2\pi} \int_0^\infty d\alpha d\beta \frac{\delta(\alpha + \beta - k_1 - k_2)}{(\alpha - k_1)^2} \left[f_{b_+ b_+ b_+}(\alpha, \beta, k_3) - f_{b_+ b_+ b_+}(k_1, k_2, k_3) \right]
$$
$$
\frac{1}{N^{3/2}} \text{Tr} \left[b_+^\dagger(\alpha) b_+^\dagger(\beta) b_+^\dagger(k_3) \right] | 0 \rangle
$$

$$+\frac{g^2N}{2\pi}\int_0^\infty d\alpha d\beta d\gamma \sum_h \frac{\delta(\alpha+\beta+\gamma-k_1)}{(\alpha+\beta)^2} f_{b_+b_+b_+}(\alpha+\beta+\gamma,k_2,k_3)\frac{1}{N^{5/2}}\mathrm{Tr}$$

$$\left[\{b_h^\dagger(\alpha),d_{-h}^\dagger(\beta)\}b_+^\dagger(\gamma)b_+^\dagger(k_2)b_+^\dagger(k_3) - \{b_h^\dagger(\alpha),d_{-h}^\dagger(\beta)\}b_+^\dagger(k_2)b_+^\dagger(k_3)b_+^\dagger(\gamma)\right]\mid 0\rangle$$

$$+\frac{g^2N}{4\pi}\int_0^\infty d\alpha d\beta d\gamma \sum_h \frac{\delta(\alpha+\beta+\gamma-k_1)}{\sqrt{\alpha\beta}(\alpha+\beta)^2} f_{b_+b_+b_+}(\alpha+\beta+\gamma,k_2,k_3)\frac{1}{N^{5/2}}\mathrm{Tr}$$

$$\left[[a_h^\dagger(\alpha),a_{-h}^\dagger(\beta)]b_+^\dagger(\gamma)b_+^\dagger(k_2)b_+^\dagger(k_3) - [a_h^\dagger(\alpha),a_{-h}^\dagger(\beta)]b_+^\dagger(k_2)b_+^\dagger(k_3)b_+^\dagger(\gamma)\right]\mid 0\rangle$$

$$+\ \text{cyclic permutations}\ \bigg\} \tag{23}$$

The five-parton states above correspond to virtual fermion-antifermion and boson-boson pair creation. The expression (23) vanishes if the wavefunction $f_{b_+b_+b_+}$ is constant.

CONCLUTIONS

We have presented a non-perturbative Hamiltonian formulation of a class of 1+1 dimensional matrix field theories, which may be derived from a classical dimensional reduction of QCD_{3+1} coupled to Dirac adjoint fermions. We choose to adopt the light-cone gauge $A_- = 0$, and are able to solve numerically the boundstate integral equations in the large-N limit. Different states may be classified according to total helicity h, and the quantum number $\mathcal{N}$, which defines the number of fermions minus the number of antifermions in a state.

For a special choice of couplings that eliminates all interactions except those involving the longitudinal current J^+ and the ϕ^4 interactions we find an infinite number of pure massless states of arbitrary length. The wavefunctions of these states are always constant, and may be solved for exactly and an example was explicitly given. In general, a massless solution involves several (possibly different) constant wavefunctions. The massless solutions observed in studies of $1+1$ dimensional supersymmetric field theories [13] are not analogous to the constant wavefunction solutions found here.

When one includes the Yukawa interactions, singularities at vanishing longitudinal momenta arise, and we show in a simple case how these are canceled by the boson and fermion self-energies. This cancellation relies on the derivation of certain 'ladder relations', which relate different wavefunctions at vanishing longitudinal momenta. These relations become singular for vanishing fermion mass m, and so in the context of the numerical techniques employed here, one is prevented from studying the limit $m \to 0$. Analytical techniques which are currently under investigation are expected to be relevant in this limiting case [8].

A particularly important property of these models is that virtual pair creation and annihilation of bosons and fermions is not suppressed in the large-N limit, and so our results go beyond the valence quark (or quenched) approximation. This provides the scope for strictly field-theoretic investigations of the internal structure of boundstates where 'sea-quarks' and small-x gluons are expected to contribute significantly to the overall polarization of a boundstate.

The techniques employed here are not specific to the choice of field theory, and are expected to have a wide range of applicability, particularly in the light-cone Hamiltonian formulation of supersymmetric field theories.

ACKNOWLEDGMENTS

This work was done in collaboration with Francesco Antonuccio. The work was supported in part by a grant from the US Department of Energy. Travel support was provided in part by a NATO collaborative grant.

References

[1] F. Antonuccio and S. Pinsky " Matrix Theory for Reduced $SU(N)$ Yang Mills with Adjoint Fermions" hep-th/9612021 to appear phys. Lett. B

[2] F. Antonuccio and S. Dalley, Phys. Lett. **B376** (1996) 154-162.

[3] F. Antonuccio and S. Dalley, Nucl. Phys. **B461** (1996) 275-301.

[4] M. Burkardt and B.van de Sande, hep-th/9510104

[5] P. A. M. Dirac, Rev. Mod. Phys. **21**, 392 (1949).

[6] S.S. Pinsky and D. G. Robertson " Light-Front QCD_{1+1} Coupled to Chiral Adjoint Fermions Phys. Lett. B379 (1996) 169 ; S. Pinsky, and R. Mohr " The Condensate for $SU(2)$ Yang-Maills Theory in 1+1 Dimensions Coupled to Massless Ajdoint Fermions" to appear in Int. J. Mod. Phys. ; S.S. Pinsky "(1+1)-Dimensional Yang-Maills Theory Coupled to Adjoint Fermions on the Light Front" hep-th/9612073

[7] S. Dalley and I.R. Klebanov, Phys. Rev. **D47** (1993) 2517; K. Demeterfi, I. R. Klebanov, and G. Bhanot, Nucl. Phys. **B418**, 15 (1994); G. Bhanot, K. Demeterfi, and I. R. Klebanov, Phys. Rev. **D48**, 4980 (1993);

[8] F. Antonuccio, S.J. Brodsky and S.Dalley, in preparation.

[9] C.M. Bender, S. Pinsky, B. Van de Sande, Phys. Rev. **D48** (1993) 816;

S. S. Pinsky, and B. van de Sande, Phys. Rev. D **49**, 2001 (1994) and J. Hiller, S.S. Pinsky and B. van de Sande, Phys.Rev. **D51** 726 (1995).

[10] G. 't Hooft, Nucl. Phys. **B75**, 461 (1974).

[11] H.-C. Pauli and S.J. Brodsky, Phys. Rev. **D32** (1985) 1993 and 2001.

[12] K. Demeterfi and I.R. Klebanov, Phys. Rev. **D48** (1993) 4980; S. Dalley and I.R. Klebanov, Phys. Lett. **B298** (1993) 79; S. Dalley, Phys. Lett. **B334** (1994) 61. F. Antonuccio and S. Dalley, Phys. Lett. **B348** (1995) 55-62.

[13] Y. Matsumura and N. Sakai, "Mass Spectra of Supersymmetric Yang-Mills Theories in $1 + 1$ dimensions", TIT/HEP-290, hep-th/9504150.

[14] A. Hashimoto and I.R. Klebanov, "Matrix Model Approach to $d > 2$ Non-critical Superstrings", PUPT-1551, hep-th/9507062.

NONPERTURBATIVE RENORMALIZATION IN LIGHT-CONE QUANTIZATION

John R. Hiller

Department of Physics
University of Minnesota, Duluth
Duluth, MN 55812

INTRODUCTION

Light-cone quantization[1] has attracted some interest as a means to perform nonperturbative analyses of quantum field theories.[2] There are good reasons to hope that this technique will provide the leverage needed to obtain a qualitative, and perhaps quantitative, connection between quantum chromodynamics (QCD) and the constituent quark model.[3] Given the complexity of QCD, it is useful to first study simpler theories such as quantum electrodynamics (QED) and even models in $1+1$ spacetime dimensions rather than $3+1$ dimensions.

Bound-state calculations in QCD_{3+1} and QED_{3+1} require nonperturbative renormalization. Most attempts at such calculations have used Tamm–Dancoff truncations[4] and cutoff-type regularization, which require counterterms that depend on Fock sector.[5] An example of such a calculation is given here for the electron's anomalous moment.[6] We then explore the practicality of Pauli–Villars regularization[7] as an alternative. In particular, we consider a simple heavy-fermion model abstracted from the Yukawa model.

We define light-cone coordinates by

$$x^{\pm} = t \pm z, \quad \mathbf{x}_{\perp} = (x, y).$$ (1)

Momentum variables are similarly constructed as

$$p^{\pm} = E \pm p_z, \quad \mathbf{p}_{\perp} = (p_x, p_y).$$ (2)

The dot product is written

$$p \cdot x = \frac{1}{2}(p^+ x^- + p^- x^+) - \mathbf{p}_{\perp} \cdot \mathbf{x}_{\perp}.$$ (3)

The time variable is taken to be x^+, and time evolution of a system is then determined by the conjugate operator $\mathcal{P}^-$. The energy E is replaced by the light-cone energy p^-, and the momentum p by the light-cone momentum $\underline{p} \equiv (p^+, \mathbf{p}_{\perp})$. The light-cone Hamiltonian is

$$H_{LC} = \mathcal{P}^+ \mathcal{P}^- - \mathcal{P}_{\perp}^2,$$ (4)

High Energy Physics and Cosmology
Edited by B.N. Kursunoglu *et al.*, Plenum Press, New York, 1997

where $\mathcal{P}^+$ and $\mathcal{P}_\perp$ are momentum operators conjugate to x^- and $\mathbf{x}_\perp$. The eigenvalue problem is

$$H_{\mathrm{LC}}\Psi = M^2\Psi, \quad \mathcal{P}\Psi = \mathit{P}\Psi, \tag{5}$$

where M is the mass of the state.

Some of the advantages of light-cone coordinates are the following: They admit the largest possible set of nondynamical generators. In particular, boosts are kinematical. For many theories of massive particles, the perturbative vacuum is the physical vacuum, because $p_i^+ = \sqrt{p^2 + m^2} + p_z > 0$ implies that no particle state can contribute to the $P^+ = 0$ vacuum. Thus there is no need to compute the vacuum state before computing massive states. Also, well-defined Fock-state expansions exist, with no disconnected vacuum pieces.

Such expansions are written as

$$\Psi = \sum_n \int [dx]_n \, [d^2k_\perp]_n \, \psi_n(x, \mathbf{k}_\perp) |n : xP^+, x\mathbf{P}_\perp + \mathbf{k}_\perp\rangle, \tag{6}$$

with n the number of particles, i ranging between 1 and n, $(P^+, \mathbf{P}_\perp)$ the total light-cone momentum, and

$$[dx]_n = 4\pi\delta\left(1 - \sum_{i=1}^n x_i\right)\prod_{i=1}^n \frac{dx_i}{4\pi\sqrt{x_i}}, \quad [d^2k_\perp]_n = 4\pi^2\delta\left(\sum_{i=1}^n \mathbf{k}_{\perp i}\right)\prod_{i=1}^n \frac{d^2k_{\perp i}}{4\pi^2}. \tag{7}$$

In the Fock basis $\{|n : p_i^+, \mathbf{p}_{\perp i}\rangle\}$, $\mathcal{P}^+$ and $\mathcal{P}_\perp$ are diagonal. The amplitude ψ_n is interpreted as the wave function of the contribution from states with n particles.

A common numerical technique is discretized light-cone quantization (DLCQ),[8] in which periodic boundary conditions are assigned to bosons and antiperiodic to fermions in a light-cone box $-L < x^- < L$, $-L_\perp < x, y < L_\perp$. Integrals are replaced by trapezoidal approximations on a grid: $p^+ \to \frac{\pi}{L}n$, $\mathbf{p}_\perp \to (\frac{\pi}{L_\perp}n_x, \frac{\pi}{L_\perp}n_y)$, with n even for bosons and odd for fermions. The limit $L \to \infty$ can be exchanged for a limit in terms of the integer *resolution* $K \equiv \frac{L}{\pi}P^+$. The longitudinal momentum fraction $x = p^+/P^+$ becomes n/K. H_{LC} is independent of L.

Because the n_i are all positive, DLCQ automatically limits the number of particles to no more than $\sim K/2$. The integers n_x and n_y range between limits associated with some maximum integer $N_\perp$ fixed by $L_\perp$ and a cutoff that limits transverse momentum.

To reduce the size of the discrete matrix problem, a Tamm–Dancoff truncation[4] in the number of particles can be applied. This has serious implications for renormalization. These include severe sector dependence of counterterms,[5] and, for QED, violation of the Ward identity.

Regularization via cutoffs typically involves limits on the invariant mass. A limit can be placed on the total invariant mass of each Fock state

$$\sum_i \frac{m_i^2 + k_{\perp i}^2}{x_i} \leq \Lambda^2 \tag{8}$$

or on the invariant mass of each particle

$$\frac{m_i^2 + k_{\perp i}^2}{x_i} \leq \Lambda^2. \tag{9}$$

There can also be a limit on the change in invariant mass across each matrix element of H_{LC}[9]

$$\left| \sum_i^n \frac{m_i^2 + k_{\perp i}^2}{x_i} - \sum_j^m \frac{m_j^2 + k_{\perp j}^2}{x_j} \right| \leq \Lambda^2. \tag{10}$$

THE ANOMALOUS MOMENT

The anomalous moment $a_e = F_2(0)$ can be computed from a spin-flip matrix element of the electromagnetic current

$$-\frac{q_1}{2m_e}F_2(q^2) = \frac{1}{2P^+}\langle P+q,\uparrow|J^+(0)|P,\downarrow\rangle \tag{11}$$

in the standard light-cone frame $q = (0, q_\perp^2/P^+, \mathbf{q}_\perp = q_1\hat{x})$. Brodsky and Drell[10] have given a useful reduction of this matrix element to the form

$$a_e = -2m_e \sum_j e_j \sum_n \int [dx]_n [d^2k_\perp]_n \, \psi_{n\uparrow}^*(x,\mathbf{k}_\perp) \sum_{i\neq j} x_i \frac{\partial}{\partial k_{1i}} \psi_{n\downarrow}(x,\mathbf{k}_\perp), \tag{12}$$

where e_j is the fractional charge of the struck constituent and $x_i = p_i^+/P^+$. The wave functions ψ_n satisfy coupled integral equations obtained from $H_{LC}\Psi = M^2\Psi$. The QED light-cone Hamiltonian has been given by Tang et al.[11] However, the bare masses and couplings must be computed from sector dependent renormalization conditions.

Consider the case where there are at most two photons and only one electron. The Fock-state expansion can be written schematically as

$$\Psi = \psi_0|e\rangle + \vec{\psi}_1|e\gamma\rangle + \vec{\psi}_2|e\gamma\gamma\rangle. \tag{13}$$

Here $\vec{\psi}_1$ and $\vec{\psi}_2$ are column vectors that contain the amplitudes for individual Fock states with one and two photons, respectively. The eigenvalue problem becomes a coupled set of three integral equations

$$\begin{aligned}
m_0^2\psi_0 + \mathbf{b}_1^\dagger \cdot \vec{\psi}_1 + \mathbf{b}_2^\dagger \cdot \vec{\psi}_2 &= M^2\psi_0, \\
\mathbf{b}_1\psi_0 + A_{11}\vec{\psi}_1 + A_{12}\vec{\psi}_2 &= M^2\vec{\psi}_1, \\
\mathbf{b}_2\psi_0 + A_{12}^\dagger\vec{\psi}_1 + A_{22}\vec{\psi}_2 &= M^2\vec{\psi}_2,
\end{aligned} \tag{14}$$

where m_0 is the bare electron mass and $\mathbf{b}_i^\dagger$ and A_{ij} are integral operators obtained from matrix elements of H_{LC}.

The bare electron mass in the one-photon sector is computed from the one-loop self energy allowed by the two-photon states. We then require that m_0 be such that $M^2 = m_e^2$ is an eigenvalue. The second and third equations can be solved for $\vec{\psi}_1/\psi_0$ and $\vec{\psi}_2/\psi_0$. Then the first equation yields m_0. Normalization of Ψ fixes the value of ψ_0.

The bare coupling for the electron-photon three-point vertex depends on the initial and final momenta of the electron and on the sectors between which the coupling acts. The momentum dependence is present because the amount of momentum available constrains the extent to which higher order corrections can contribute. Similarly, the sector dependence makes itself felt when the number of additional particles in higher-order corrections is restricted. The coupling is fixed by the ratio of the $e\gamma \to e$ transition matrix element to the bare vertex at zero photon momentum.

In the present calculation we use a Tamm–Dancoff truncation to $\{e, e\gamma, e\gamma\gamma\}$, a nonzero photon mass $m_\gamma = m_e/10$, and a moderate coupling $\alpha = 1/10$. Some results are given elsewhere.[6] When only states with at most one photon and no pairs are retained, one can show that a_e reduces to

$$a_e = \frac{\alpha m_e^2}{\pi^2} \int \frac{dx\, d^2k_\perp}{x} \frac{\theta(\Lambda^2 - (m_e^2+k_\perp^2)/x - (m_\gamma^2+k_\perp^2)/(1-x))}{[m_e^2 - (m_e^2+k_\perp^2)/x - (m_\gamma^2+k_\perp^2)/(1-x)]^2}, \tag{15}$$

which in the limit of $\Lambda \longrightarrow \infty$ becomes[10]

$$a_e = \frac{\alpha}{2\pi}\int_0^1 \frac{2x^2(1-x)dx}{x^2 + (1-x)(m_\gamma/m_e)^2}. \tag{16}$$

For $m_\gamma = 0$, this yields the standard Schwinger contribution[12] of $\alpha/2\pi$.

195

Table 1. Values of the subtracted

integral $I_{sub}(M^2/\mu^2, \mu_i^2/\mu^2)$ in the
limit of infinite cutoff. The Pauli–
Villars masses are $\mu_1^2 = 10\mu^2$,
$\mu_2^2 = 50\mu^2$ and $\mu_3^2 = 100\mu^2$.

M^2	0	$0.05\mu^2$	$0.1\mu^2$	$0.2\mu^2$
I_{sub}	-0.064	0.70	1.37	2.70

YUKAWA THEORY AT ONE LOOP

As an alternative approach to regularization, we consider Pauli–Villars[7] regularization of the $3 + 1$ Yukawa model.[13, 14] The one-loop fermion self-energy is proportional to

$$I(\mu^2, M^2) \equiv -\frac{1}{\mu^2} \int \frac{dl^+ d^2 l_\perp}{l^+ (q^+ - l^+)^2} \frac{(q^+)^2 l_\perp^2 + (2q^+ - l^+)^2 M^2}{M^2 - D_1} \theta(\Lambda^2 - D_1), \quad (17)$$

where q is the fermion momentum, μ is the boson mass, M is the fermion mass, and

$$D_1 = \frac{\mu^2 + l_\perp^2}{l^+/q^+} + \frac{M^2 + l_\perp^2}{(q^+ - l^+)/q^+}. \quad (18)$$

The boson mass μ sets the energy scale. When $M^2 = 0$ we obtain

$$I(\mu^2, 0) = \frac{\pi}{\mu^2} \left[\frac{\Lambda^2}{2} - \frac{\mu^4}{2\Lambda^2} - \mu^2 \ln\left(\frac{\Lambda^2}{\mu^2}\right) \right]. \quad (19)$$

In order to maintain $I(\mu^2, M^2) \propto M^2$, three Pauli-Villars bosons are needed:[15]

$$I_{sub}(\mu^2, M^2, \mu_i^2) = I(\mu^2, M^2) + \sum_{i=1}^{3} C_i I(\mu_i^2, M^2). \quad (20)$$

The C_i are chosen to satisfy

$$1 + \sum_{i=1}^{3} C_i = 0, \quad \mu^2 + \sum_{i=1}^{3} C_i \mu_i^2 = 0, \quad \sum_{i=1}^{3} C_i \mu_i^2 \ln(\mu_i^2/\mu^2) = 0. \quad (21)$$

A DLCQ calculation of I_{sub} has been done,[16] with values of 20, 22, and 24 for K and 25 through 30 for $N_\perp$. Modification of the trapezoidal rule, with introduction of unequal weights, is necessary to obtain sufficient accuracy. Each integral in (20) was separately extrapolated to infinite K and $N_\perp$ via fits to either $c_0 + a_1/K^3 + b_1/N_\perp^2$ or $c_0 + a_1/K^3 + a_2/K^4 + b_1/N_\perp^2 + b_2/N_\perp^3$. The latter was used for the μ_1 integral. Extrapolation after subtraction is not as accurate. The resulting values of I_{sub} were extrapolated to infinite cutoff by fits to $a + b/\Lambda^2$. These fully extrapolated values are given in Table 1.

The magnitude of the error in each extrapolated integral was found to be ≤ 0.02 when compared to the analytic result for $M^2 = 0$. This implies an error of ± 0.04 in the I_{sub} values. The extrapolation in Λ^2 induces additional uncertainty reflected in the miss of zero by 0.06 for $M^2 = 0$. The values in Table 1 are consistent with $I_{sub} \propto M^2$ to within this amount of error.

The number of Fock states required for Pauli–Villars particles is approximately 1.5 times the number for physical states. A listing of counts for two cases is given in Table 2. Making μ_1 larger does decrease the number of Pauli-Villars states but this increases the

Table 2. Number of Fock states used in two typical cases.

Λ^2/μ^2	K	$N_\perp$	physical boson states	Pauli-Villars boson states			
				$\mu_1^2 = 10\mu^2$	$\mu_2^2 = 50\mu^2$	$\mu_3^2 = 100\mu^2$	total
200	20	25	25975	22602	11142	3305	37049
200	24	30	44943	39162	19293	5695	64150

coefficients C_i and thereby amplifies errors in the integrals. Also, with fewer states, the integrals themselves are approximated less accurately.

We could also consider the boson self energy. To lowest order there is a fermion loop contribution

$$\int \frac{dl^+ d^2 l_\perp}{4LL_\perp^2} \frac{q^+(l_\perp^2 + M^2)}{l^{+2}(q^+ - l^+)^2} \frac{\theta(\Lambda^2 - D_2)}{\mu^2 - D_2}, \tag{22}$$

where

$$D_2 \equiv q^{+2}(M^2 + l_\perp^2)/[l^+(q^+ - l^+)], \tag{23}$$

and a ϕ^4 contribution

$$\int \frac{dl^+ d^2 l_\perp dk^+ d^2 k_\perp}{q^+ l^+ k^+ (q^+ - l^+ - k^+)} \frac{\theta(\Lambda^2 - D_4)}{\mu^2 - D_4}, \tag{24}$$

where

$$D_4 \equiv \frac{\mu^2 + l_\perp^2}{l^+/q^+} + \frac{\mu^2 + k_\perp^2}{k^+/q^+} + \frac{\mu^2 + (l_\perp + k_\perp)^2}{(q^+ - l^+ - k^+)/q^+}. \tag{25}$$

A Pauli–Villars fermion may be needed.

A HEAVY-FERMION MODEL

By some severe modifications of the Yukawa Hamiltonian[17] we obtain the following model Hamiltonian:

$$H_{LC}^{\text{eff}} = M_0^2 \int \frac{dp^+ d^2 p_\perp}{16\pi^3 p^+} \sum_\sigma b_{p\sigma}^\dagger b_{p\sigma} + P^+ \int \frac{dq^+ d^2 q_\perp}{16\pi^3 q^+} \left[\frac{\mu^2 + q_\perp^2}{q^+} a_q^\dagger a_q + \frac{\mu_1^2 + q_\perp^2}{q^+} a_{1q}^\dagger a_{1q} \right]$$

$$+ g \int \frac{dp_1^+ d^2 p_{\perp 1}}{\sqrt{16\pi^3 p_1^+}} \int \frac{dp_2^+ d^2 p_{\perp 2}}{\sqrt{16\pi^3 p_2^+}} \int \frac{dq^+ d^2 q_\perp}{16\pi^3 q^+} \sum_\sigma b_{p_1\sigma}^\dagger b_{p_2\sigma} \tag{26}$$

$$\times \left[a_q^\dagger \delta(p_1 - p_2 + q) + a_q \delta(p_1 - p_2 - q) \right.$$

$$\left. + i a_{1q}^\dagger \delta(p_1 - p_2 + q) + i a_{1q} \delta(p_1 - p_2 - q) \right].$$

The kinetic energy of the fermion is no longer momentum dependent and only a modified no-flip three-point vertex remains as an interaction. The fermion then acts as a "static" source for the boson. We include one Pauli–Villars field, which will prove sufficient in this case. Similar Hamiltonians, without the Pauli–Villars field, have been considered in equal-time[18] and light-cone coordinates.[19]

We write the eigenvector as a Fock-state expansion

$$\Phi_\sigma = \sqrt{16\pi^3 P^+} \sum_{n,n_1} \int \frac{dp^+ d^2 p_\perp}{\sqrt{16\pi^3 p^+}} \prod_{i=1}^n \int \frac{dq_i^+ d^2 q_{\perp i}}{\sqrt{16\pi^3 q_i^+}} \prod_{j=1}^{n_1} \int \frac{dr_j^+ d^2 r_{\perp j}}{\sqrt{16\pi^3 r_j^+}} \tag{27}$$

$$\times \delta(P - p - \sum_i^n q_i - \sum_j^{n_1} r_j) \phi^{(n,n_1)}(q_i, r_j; p) \frac{1}{\sqrt{n! n_1!}} b_{p\sigma}^\dagger \prod_i^n a_{q_i}^\dagger \prod_j^{n_1} a_{1r_j}^\dagger |0\rangle,$$

normalized according to $\Phi_\sigma'^\dagger \cdot \Phi_\sigma = 16\pi^3 P^+ \delta(\underline{P}' - \underline{P})$, which yields

$$1 = \sum_{n,n_1} \prod_i^n \int dq_i^+ d^2 q_{\perp i} \prod_j^{n_1} \int dr_j^+ d^2 r_{\perp j} \left| \phi^{(n,n_1)}(\underline{q}_i, \underline{r}_j; \underline{P} - \sum_i \underline{q}_i - \sum_j \underline{r}_j) \right|^2 . \tag{28}$$

For Φ_σ to satisfy the Schrödinger equation (5), the amplitudes must satisfy

$$\left[M^2 - M_0^2 \quad - \sum_i \frac{\mu^2 + q_{\perp i}^2}{y_i} - \sum_j \frac{\mu_1^2 + r_{\perp j}^2}{z_j} \right] \phi^{(n,n_1)} \tag{29}$$

$$= g \left\{ \sqrt{n+1} \int \frac{dq^+ d^2 q_\perp}{\sqrt{16\pi^3 q^+}} \phi^{(n+1,n_1)}(\underline{q}_i, \underline{q}, \underline{r}_j, \underline{p}) \right.$$

$$+ \frac{1}{\sqrt{n}} \sum_i \frac{1}{\sqrt{16\pi^3 q_i^+}} \phi^{(n-1,n_1)}(\underline{q}_1, \ldots, \underline{q}_{i-1}, \underline{q}_{i+1}, \ldots, \underline{q}_n, \underline{r}_j, \underline{p})$$

$$+ i\sqrt{n_1+1} \int \frac{dr^+ d^2 r_\perp}{\sqrt{16\pi^3 r^+}} \phi^{(n,n_1+1)}(\underline{q}_i, \underline{r}_j, \underline{r}, \underline{p})$$

$$\left. + \frac{i}{\sqrt{n}} \sum_j \frac{1}{\sqrt{16\pi^3 r_j^+}} \phi^{(n,n_1-1)}(\underline{q}_i, \underline{r}_1, \ldots, \underline{r}_{j-1}, \underline{r}_{j+1}, \ldots, \underline{r}_{n_1}, \underline{p}) \right\} .$$

The structure of this coupled set of integral equations is deliberately identical in basic form to the equations considered by Greenberg and Schweber.[18] Therefore, we transcribe their *ansatz* for a solution to light-cone form

$$\phi^{(n,n_1)} = \sqrt{Z} \frac{(-g)^n (-ig)^{n_1}}{\sqrt{n! n_1!}} \prod_i \frac{q_i^+}{\sqrt{16\pi^3 q_i^+}(\mu^2 + q_{\perp i}^2)} \prod_j \frac{r_j^+}{\sqrt{16\pi^3 r_j^+}(\mu_1^2 + r_{\perp j}^2)} . \tag{30}$$

This does work as a solution if $M_0^2(\mu_1)$ is chosen to satisfy

$$M^2 - M_0^2 = -\frac{g^2}{16\pi^3} \left\{ \int \frac{dy d^2 q_\perp}{\mu^2 + q_\perp^2} - \int \frac{dz d^2 r_\perp}{\mu_1^2 + r_\perp^2} \right\} . \tag{31}$$

From the normalization condition (28) we obtain

$$\frac{1}{Z} = \exp \left\{ \frac{g^2}{16\pi^3} \left[\int \frac{y dy d^2 q_\perp}{(\mu^2 + q_\perp^2)^2} + \int \frac{z dz d^2 r_\perp}{(\mu_1^2 + r_\perp^2)^2} \right] \right\} . \tag{32}$$

The bare mass and wave function renormalization are thus determined as functions of the Pauli–Villars mass.

To fix the coupling we could use the slope of the fermion no-flip form factor, which is related to the transverse size of the dressed fermion. The form factor is most easily evaluated from[10]

$$F(Q^2) = \frac{1}{2P^+} \langle P + p_\gamma \uparrow | J^+(0) | P \uparrow \rangle \tag{33}$$

$$= \sum_j e_j \int 16\pi^3 \delta(1 - \sum_i x_i) \delta(\sum_i \mathbf{k}_{\perp i}) \prod_i \frac{dx_i d^2 p_{\perp i}}{16\pi^3}$$

$$\times \psi^*_{P+p_\gamma \uparrow}(x_i, \mathbf{p}'_{\perp i}) \psi_{P\uparrow}(x_i, \mathbf{p}_{\perp i}),$$

where the matrix element has been evaluated in the frame with

$$P = (P^+, P^- = \frac{M^2}{P^+}, \mathbf{0}_\perp), \quad p_\gamma = (0, p_\gamma^- = 2p_\gamma \cdot P/P^+, \mathbf{p}_{\gamma \perp}), \quad Q^2 \equiv p_{\gamma \perp}^2 , \tag{34}$$

e_j is the charge of the jth constituent, and

$$\mathbf{p}'_{\perp i} = \begin{cases} \mathbf{p}_{\perp i} - x_i \mathbf{p}_{\gamma \perp} & i \neq j \\ \mathbf{p}_{\perp i} + (1 - x_i)\mathbf{p}_{\gamma \perp} & i = j. \end{cases} \tag{35}$$

A sum over Fock states is understood.

When the fermion is assigned a charge of 1, and the bosons remain neutral, the analytic solution for the amplitudes yields

$$F(Q^2) = Z \exp\left\{ g^2 \int \frac{dy\, d^2 q_\perp}{16\pi^3} \frac{\sqrt{y}}{\mu^2 + q_\perp^2} \frac{\sqrt{y}}{\mu^2 + q_\perp'^2} + \text{P-V term} \right\}, \tag{36}$$

with

$$\mathbf{q}'_\perp = \mathbf{q}_\perp - y \mathbf{p}_{\gamma \perp}. \tag{37}$$

From this we find

$$F'(0) = -g^2 \int \frac{dy\, d^2 q_\perp}{16\pi^3} \frac{y^3}{(\mu^2 + q_\perp^2)^3} \left[\frac{2\mu^2}{\mu^2 + q_\perp^2} - 1 \right] + \text{P-V term}. \tag{38}$$

Numerically, the slope is computed from a finite-difference approximation to

$$F'(0) = \sum_{n,n_1} \prod_i^n \int dq_i^+ d^2 q_{\perp i} \prod_j^{n_1} \int dr_j^+ d^2 r_{\perp j} \tag{39}$$

$$\times \left[\left(\sum_i \frac{y_i^2}{4} \nabla_{\perp i}^2 + \sum_j \frac{z_j^2}{4} \nabla_{\perp j}^2 \right) \phi^{(n,n_1)}\left(q_i, r_j; P - \sum_i q_i - \sum_j r_j\right) \right]^*$$

$$\times \phi^{(n,n_1)}\left(q_i, r_j; P - \sum_i q_i - \sum_j r_j\right).$$

With the bare parameters determined, we "predict" a value for $\langle : \phi^2(0) : \rangle$. For the analytic solution, this expectation value reduces to

$$\langle : \phi^2(0) : \rangle = \frac{g^2}{8\pi^2 \mu^2} \left[1 - \frac{\mu^2}{\Lambda^2} - \frac{\mu^2}{\Lambda^2} \ln \frac{\mu^2}{\Lambda^2} \right]. \tag{40}$$

From a numerical solution it can be computed from a sum similar to the normalization sum

$$\langle : \phi^2(0) : \rangle = \sum_{n=1, n_1 = 0} \prod_i^n \int dq_i^+ d^2 q_{\perp i} \prod_j^{n_1} \int dr_j^+ d^2 r_{\perp j} \left(\sum_{k=1}^n \frac{2}{q_k^+ / P^+} \right) \tag{41}$$

$$\times \left| \phi^{(n,n_1)}\left(q_i, r_j; P - \sum_i q_i - \sum_j r_j\right) \right|^2.$$

SUMMARY

For the anomalous moment calculation there remain several hurdles. The Tamm–Dancoff truncation results in logarithmically divergent four-point graphs. To deal with these will probably require use of scattering processes, such as Compton scattering,[20] to obtain renormalization conditions. Verification of the removal of all logarithms and restoration of symmetries can then be undertaken. Also neglected up to this point have been zero modes, photon modes of zero longitudinal momentum.[21] How they might be included has been indicated by Kalloniatis and Robertson.[22]

Additional physics could be included in the calculation by introducing an effective interaction from Z graphs or even putting eee^+ states in the basis. In the latter case, photon mass renormalization must be done.

In the Yukawa-model calculations we have learned that Pauli–Villars Fock states increase the basis size by only 150%, which may not be prohibitive. To perform such calculations accurately with a minimal basis size, improvement of ordinary DLCQ, by inclusion of weighting factors, is critical.

We have found a simple $3 + 1$ model, related to Yukawa theory, which can be solved analytically. Here we will attempt a nonperturbative numerical solution to further test the use of Pauli–Villars regularization in DLCQ. If successful, we can begin to increase the complexity of the model, eventually reaching the full Yukawa theory.

ACKNOWLEDGMENTS

This work was supported in part by the Minnesota Supercomputer Institute through grants of computing time. It was done in collaboration with S.J. Brodsky and G. McCartor.

REFERENCES

1. P.A.M. Dirac, *Rev. Mod. Phys.* 21:392 (1949).
2. For reviews and references to recent work, see S.J. Brodsky and H.-C. Pauli, in *Recent Aspects of Quantum Fields*, Lecture Notes in Physics Vol. 396, H. Mitter and H. Gausterer, eds., Springer-Verlag, Berlin (1991); S.J. Brodsky, G. McCartor, H.-C. Pauli, and S.S. Pinsky, *Part. World* 3:109 (1993); M. Burkardt, *Adv. Nucl. Phys.* 23:1 (1996).
3. K.G. Wilson, T.S. Walhout, A. Harindranath, W.-M. Zhang, R.J. Perry, and St.D. G lazek, *Phys. Rev.* D49:6720 (1994).
4. I. Tamm, *J. Phys. (Moscow)* 9:449 (1945); S.M. Dancoff, *Phys. Rev.* 78:382 (1950).
5. R.J. Perry, A. Harindranath, and K.G. Wilson, *Phys. Rev. Lett.* 65:2959 (1990).
6. J.R. Hiller, in *Theory of Hadrons and Light-Front QCD*, St.D. G lazek, ed., World Scientific, Singapore (1995), p. 277; J.R. Hiller and S.J. Brodsky, Nonperturbative renormalization of QED in light-cone quantization, hep-ph/9608493, to appear in the proceedings of DPF96, Minneapolis, MN, August 11-15, 1996; J.R. Hiller and S.J. Brodsky, in preparation.
7. W. Pauli and F. Villars, *Rev. Mod. Phys.* 21:4334 (1949).
8. H.-C. Pauli and S.J. Brodsky, *Phys. Rev.* D32:1993 (1985); D32:2001 (1985).
9. G.P. Lepage, summary talk, SMU light-cone conference (1991).
10. S.J. Brodsky and S.D. Drell, *Phys. Rev.* D22:2236 (1980).
11. A.C. Tang, S.J. Brodsky, and H.-C. Pauli, *Phys. Rev.* D44:1842 (1991).
12. J. Schwinger, *Phys. Rev.* 73:416 (1948); 76:790 (1949).
13. For a light-cone Tamm–Dancoff bound-state calculation, see St. G lazek, A. Harindranath, S. Pinsky, J. Shigemitsu, and K. Wilson, *Phys. Rev.* D47:1599 (1993); P.M. Wort, *ibid.* 47:608 (1993).
14. For a discussion of Pauli–Villars regularization in the context of light-cone quantization and the Yukawa model, see M. Burkardt and A. Langnau, *Phys. Rev.* D44:1187 (1991).
15. C. Bouchiat, P. Fayet, and N. Sourlas, *Lett. Nuovo Cim.* 4:9 (1972); S.-J. Chang and T.-M. Yan, *Phys. Rev.* D7:1147 (1973).
16. S.J. Brodsky, J.R. Hiller, and G. McCartor, in preparation.
17. G. McCartor and D.G. Robertson, *Z. Phys.* C53:679 (1992).
18. O.W. Greenberg and S.S. Schweber, *Nuovo Cim.* 8:378 (1958); S.S. Schweber, *An introduction to relativistic quantum field theory*, Row, Peterson, Evanston, IL (1961), p. 339.
19. St.D. G lazek and R.J. Perry, *Phys. Rev.* D45:3734 (1992).
20. D.Mustaki and S. Pinsky, *Phys. Rev.* D45:3775 (1992).
21. See, for example, A.C. Kalloniatis and H.-C. Pauli, *Z. Phys.* C60:255 (1993); 63:161 (1994).
22. A.C. Kalloniatis and D.G. Robertson, *Phys. Rev.* D50:5262 (1994).

SINGLET-TRIPLET SPLITTING OF POSITRONIUM IN LIGHT-FRONT QED

Billy D. Jones

Department of Physics
The Ohio State University
Columbus OH 43210

INTRODUCTION

We study the QED bound-state problem in a light-front hamiltonian approach. It is important to establish the equivalence (or not) of equal-time and light-front approaches in the well-understood arena of Quantum Electrodynamics. Along these lines, the singlet-triplet ground state spin splitting in positronium is calculated. The well-known result, $\frac{7}{6}\alpha^2 Ryd$, is obtained analytically, which establishes the equivalence between the equal-time and light-front approaches (at least to this order). The true equivalence of the two approaches can only be established after higher-order calculations. It was previously shown that this light-front result could be obtained analytically [1], but a simpler method is presented in this paper.

A calculation of the singlet-triplet ground state spin splitting in positronium is rather trivial from the viewpoint of a Coulomb gauge equal-time calculation (see for example §83–84 of [2]). This is not the case in a light-cone gauge light-front calculation (see for example [3]). We will briefly outline the derivation of the effective Hamiltonian, and then proceed with an analytic calculation of the singlet-triplet splitting.

EFFECTIVE HAMILTONIAN

A brief description of the approach will be given, and the resulting effective Hamiltonian which will be studied in this paper will be written. For details of the derivation of the effective Hamiltonian, and for the original references, see [1].

The starting point is the canonical QED Hamiltonian in the light-cone gauge. Then a regulator is introduced, Λ, which removes high energy exchanges from the theory. Proceeding, a unitary transformation is defined that acts on the regulated canonical Hamiltonian and produces an effective Hamiltonian at a lower energy scale, λ. The transformation is unitary, so the spectrum of the regulated canonical Hamiltonian and the effective Hamiltonian are equivalent (of course approximations can invalidate this conclusion). The $\Lambda \longrightarrow \infty$ limit is studied, and the regulated canonical Hamiltonian is adjusted so that this limit can be taken. So far this is nothing but the old story of renormalization, but the procedure is far from trivial in a light-front approach since longitudinal locality is lost.[1] Unfortunately a complete story of how the renormalization works out is not available, the study is very much a work in progress. However, to obtain the results of this paper consistently, only the one loop electron self-energy renormalization needs to be performed. This will not be shown explicitly in this paper, but is in [1]. The resulting effective Hamiltonian satisfies the Schrödinger equation written below, which is conveniently written in the notation of one body Quantum Mechanics. Note, to obtain this form of the second order effective Hamiltonian, we needed to place the scale in the following window, $m\alpha^2 \ll \lambda \ll m\alpha$. This lower bound is the nonperturbative energy scale of interest; if λ is lowered below this bound, the Coulomb interaction does not arise from the second order effective interactions alone. This upper bound is the dominant energy of emitted and absorbed photons; placing λ below this bound allows the leading order results to be obtained in the valence sector alone. Given this, the second order effective Hamiltonian satisfies the following Schrödinger equation

$$\left(\hat{\mathcal{H}}_o + \hat{\mathcal{V}}\right)|\Phi_N\rangle = M_N^2|\Phi_N\rangle ,$$ (1)

where M_N is the mass of the state and

- $\langle \Phi_N|\Phi_{N'}\rangle = \delta_{NN'}$ (2)
- $1 = \sum_{s_1 s_2} \int d^3p \, |\mathbf{p}s_1s_2\rangle\langle\mathbf{p}s_1s_2| = \sum_{s_1 s_2} \int d^3x \, |\mathbf{x}s_1s_2\rangle\langle\mathbf{x}s_1s_2| = \sum_N |\Phi_N\rangle\langle\Phi_N|$ (3)
- $\langle \mathbf{p}'s_3s_4|\hat{\mathcal{V}}|\mathbf{p}s_1s_2\rangle = \mathcal{V}(\mathbf{p}'s_3s_4; \mathbf{p}s_1s_2)$ (4)
- $\langle \mathbf{p}'s_3s_4|\hat{\mathcal{H}}_o|\mathbf{p}s_1s_2\rangle = 4(m^2 + \mathbf{p}^2)\delta^3(p - p')\delta_{s_1s_3}\delta_{s_2s_4} - (4m)\dfrac{\alpha}{2\pi^2}\dfrac{\delta_{s_1s_3}\delta_{s_2s_4}}{(\mathbf{p}-\mathbf{p}')^2}$ (5)
- $M_N^2 = (2m + B_N)^2 .$ (6)

m is the electron mass, $-B_N$ is the binding energy, and N labels all the quantum numbers of the state. For notational purposes note that we label the final relative three-momentum with a prime, and that the initial and final electrons are labeled by "1" and "3" respectively, and the initial and final positrons are labeled by "2" and "4" respectively. Before proceeding to write $\hat{\mathcal{V}}$, it is convenient to discuss the spectrum of $\hat{\mathcal{H}}_o$.

In zeroth order $\hat{\mathcal{V}}$ is neglected and Eq.(1) becomes

$$\hat{\mathcal{H}}_o|\phi_N\rangle = \mathcal{M}_N^2|\phi_N\rangle = \left(4m^2 + 4m\mathcal{B}_N\right)|\phi_N\rangle .$$ (7)

This last equality defines our zeroth order binding energy, $-\mathcal{B}_N$. Projecting this eigenvalue equation into momentum space gives

$$\left(-\mathcal{B}_N + \dfrac{\mathbf{p}'^2}{m}\right)\phi_N(\mathbf{p}'s_3s_4) = \dfrac{\alpha}{2\pi^2}\int \dfrac{d^3p}{(\mathbf{p}-\mathbf{p}')^2}\phi_N(\mathbf{p}s_3s_4) ,$$ (8)

---

[1]On a positive note, recall that there is an exact scale invariance of the theory under a longitudinal scaling; thus no nonperturbative longitudinal scale can arise through the process of renormalization.

the familiar non-relativistic Schrödinger equation for positronium.

After a simplification detailed in the next Section and mentioned in the Abstract, like the Coulomb gauge equal-time calculation, to obtain the ground state singlet-triplet splitting, only the wave function at the origin is required, which we thus record:

$$(\phi_N(\mathbf{x}=0))^2 = \frac{1}{(2\pi)^3}\left(\int d^3p\ \phi_N(\mathbf{p})\right)^2 = \frac{1}{\pi}\left(\frac{m\alpha}{2n}\right)^3\delta_{l,0}\,. \tag{9}$$

n is the principal quantum number, and l is the angular momentum quantum number.

Now we proceed to write $\hat{\mathcal{V}}$. Note that $\hat{\mathcal{V}}$ is not diagonal in momentum space, so if we define

$$\hat{\mathcal{V}} = \hat{\mathcal{V}}^{(0)} + \hat{\mathcal{V}}^{(1)} + \hat{\mathcal{V}}^{(2)} + \cdots, \tag{10}$$

where the superscript implies the α-scaling of a matrix element of the operator in momentum space, then in first order bound-state perturbation theory these operators contribute

$$\langle\phi_N|\hat{\mathcal{V}}^{(S)}|\phi_{N'}\rangle \sim m^2\alpha^{3+S}\,. \tag{11}$$

Note that Eq. (10) starts at $S=0$; this is a result of the derivation of the effective Hamiltonian. Interestingly, note that in a Coulomb gauge equal-time calculation, this series starts at $S=1$. In summary, to be consistent to order α^4, we need to look at all matrix elements $\mathcal{V}^{(S)}(\mathbf{p}'s_3s_4;\mathbf{p}s_1s_2)$ with $S\leq 1$.[2]

Before proceeding to write out these expressions for $\hat{\mathcal{V}}^{(S)}$, note that we only calculate spin splittings, so constants along the diagonal in spin space were neglected. Given this, to get the spin splittings correct to order α^4 we need to consider

$$\mathcal{V}^{(0)}(\mathbf{p}'s_3s_4;\mathbf{p}s_1s_2) = \frac{-c_{ex}e^2}{4\pi^3(\mathbf{p}-\mathbf{p}')^2}v^{(0)}(\mathbf{p}'s_3s_4;\mathbf{p}s_1s_2)\,, \tag{12}$$

where

$$v^{(0)}(\mathbf{p}'s_3s_4;\mathbf{p}s_1s_2) = \left(\delta_{s_1\bar{s}_3}\delta_{s_2s_4}f_1(\mathbf{p}'s_3s_4;\mathbf{p}s_1s_2) + \delta_{s_1s_3}\delta_{s_2\bar{s}_4}f_2(\mathbf{p}'s_3s_4;\mathbf{p}s_1s_2)\right) \tag{13}$$

$$f_1(\mathbf{p}'s_3s_4;\mathbf{p}s_1s_2) = s_1(p_y-p_y') - i(p_x-p_x')\,, \tag{14}$$

$$f_2(\mathbf{p}'s_3s_4;\mathbf{p}s_1s_2) = s_4(p_y-p_y') + i(p_x-p_x')\,. \tag{15}$$

$s_i/2$ is the spin quantum number of fermion "i"; $s_i=\pm1$ $(i=1,2,3,4)$ only; $\bar{s}_i=-s_i$. The only other interaction that needs to be considered is

$$\mathcal{V}^{(1)}(\mathbf{p}'s_3s_4;\mathbf{p}s_1s_2) = \frac{e^2}{4m\pi^3}\left(c_{an}\delta_{s_1s_2}\delta_{s_1s_4}\delta_{s_3s_4} + c_{ex}\delta_{s_2\bar{s}_4}\delta_{s_2\bar{s}_1}\delta_{s_3\bar{s}_1} + \right.$$
$$\left. + \left(c_{an}\frac{1}{2} - c_{ex}\frac{(p_\perp-p'_\perp)^2}{(\mathbf{p}-\mathbf{p}')^2}\right)\delta_{s_1\bar{s}_2}\delta_{s_3\bar{s}_4}\right)\,. \tag{16}$$

The constants c_{ex} and c_{an} were introduced only to distinguish the terms that arise from the 'exchange' and 'annihilation' channels respectively; $c_{ex}=c_{an}=1$.

---

[2]For example, $\frac{m e^2 \mathbf{p}}{(\mathbf{p}-\mathbf{p}')^3} \sim \frac{\alpha^2}{\alpha^3} \Longrightarrow S=0$.

SINGLET-TRIPLET SPLITTING

Now we will calculate the ground state singlet-triplet splitting to order α^4 using bound-state perturbation theory in $\hat{\mathcal{V}}$. Perhaps the most straightforward approach is to just get busy and calculate, since the non-relativistic Coulomb spectrum is so well known. This is exactly what is done in [1]; however, as can be seen by the complexity of Appendix C in that paper, the calculation is complicated and at the level of a "Lamb shift calculation." We will now present a simpler method to calculate this shift.[3] This simpler method uses a unitary transformation to "remove" $\hat{\mathcal{V}}^{(0)}$ much in the spirit of Schwinger's early QED calculations [6]. This simpler method now follows.

First, set up a general unitary transformation with hermitian generator $\hat{Q}$:

$$\hat{H} = \hat{\mathcal{H}}_o + \hat{\mathcal{V}}^{(0)} + \hat{\mathcal{V}}^{(1)} + \hat{\mathcal{V}}^{(2)} + \cdots , \tag{17}$$

$$\hat{H}' = e^{i\hat{Q}} \hat{H} e^{-i\hat{Q}}$$

$$= \hat{H} + i \left[\hat{Q}, \hat{H}\right] + \frac{i^2}{2!} \left[\hat{Q}, \left[\hat{Q}, \hat{H}\right]\right] + \cdots . \tag{18}$$

Now define $\hat{Q}$ by requiring its commutator with $\hat{\mathcal{H}}_o$ to cancel $\hat{\mathcal{V}}^{(0)}$:

$$\hat{\mathcal{V}}^{(0)} + i \left[\hat{Q}, \hat{\mathcal{H}}_o\right] = 0 . \tag{19}$$

Putting this into Eq. (18) gives

$$\hat{H}' = \hat{\mathcal{H}}_o + \left(1 - \frac{1}{2!}\right) \left[i\hat{Q}, \hat{\mathcal{V}}^{(0)}\right] + e^{i\hat{Q}} \left(\hat{\mathcal{V}}^{(1)} + \hat{\mathcal{V}}^{(2)} + \cdots\right) e^{-i\hat{Q}}$$

$$+ \left(\left(\frac{1}{2!} - \frac{1}{3!}\right) \left[i\hat{Q}, \left[i\hat{Q}, \hat{\mathcal{V}}^{(0)}\right]\right] + \left(\frac{1}{3!} - \frac{1}{4!}\right) \left[i\hat{Q}, \left[i\hat{Q}, \left[i\hat{Q}, \hat{\mathcal{V}}^{(0)}\right]\right]\right] + \cdots\right) \tag{20}$$

Note that $\hat{H}$ and $\hat{H}'$ have equivalent lowest order spectrums given by $\hat{\mathcal{H}}_o$; this can be seen easily by looking at matrix elements of the equations in Coulomb states, that is in states of $\hat{\mathcal{H}}_o$. To summarize, we must diagonalize the following interaction in spin space to obtain the order α^4 ground state singlet-triplet splitting in positronium:

$$\delta^{(1)} M^2(s_3, s_4; s_1, s_2) = \langle \phi_{1,0,0,s_3,s_4} | \hat{\mathcal{V}}^{(1)} + \frac{1}{2} \left[i\hat{Q}, \hat{\mathcal{V}}^{(0)}\right] | \phi_{1,0,0,s_1,s_2} \rangle , \tag{21}$$

where $\hat{Q}$ is a solution to Eq. (19). The superscript on $\delta^{(1)} M^2$ signifies that it is a *first* order bound-state perturbation theory shift. The quantum numbers are $N = (n, l, m_l, s_e, s_{\bar{e}}) \longrightarrow (1, 0, 0, s_e, s_{\bar{e}})$ for the ground state.

In what follows we will solve Eq. (19) for $\hat{Q}$ in the free basis in momentum space,[4] and then calculate the shift defined by Eq. (21).

From the form of $\hat{\mathcal{V}}^{(0)}$ and $\hat{\mathcal{H}}_o$ we see that $\hat{Q}$ has the following general form

$$\langle \mathbf{p}' s_3 s_4 | i\hat{Q} | \mathbf{p} s_1 s_2 \rangle = \delta^3(p - p') \langle \mathbf{p}' s_3 s_4 | i\hat{R} | \mathbf{p} s_1 s_2 \rangle , \tag{22}$$

where from Eq. (19), $\hat{R}$ satisfies

$$\frac{v^{(0)}(\mathbf{p}' s_3 s_4; \mathbf{p} s_1 s_2)}{2m} = \langle \mathbf{p} s_3 s_4 | i\hat{R} | \mathbf{p} s_1 s_2 \rangle - \langle \mathbf{p}' s_3 s_4 | i\hat{R} | \mathbf{p}' s_1 s_2 \rangle . \tag{23}$$

---

[3]The idea behind this simpler method originated with Brisudová and Perry [4].

[4]This is the trick, to solve for $\hat{Q}$ in the free basis; if $\hat{Q}$ is solved for in the Coulomb basis the calculation follows the one carried out in [1].

Recall Eq. (13) for the form of $v^{(0)}$. Thus, the general form of $\hat{R}$ is

$$\langle \mathbf{p}s_3s_4|i\hat{R}|\mathbf{p}s_1s_2\rangle = \frac{\delta_{s_1\bar{s}_3}\delta_{s_2s_4}}{2m}(s_1p_y - ip_x) + \frac{\delta_{s_1s_3}\delta_{s_2\bar{s}_4}}{2m}(s_4p_y + ip_x) \ . \qquad (24)$$

Since $\hat{Q}$ is diagonal in momentum space it is a simple matter to calculate the contributions from Eq. (21). Define

$$\delta M_1^2 = \langle\phi_{1,0,0,s_3,s_4}|\hat{\mathcal{V}}^{(1)}|\phi_{1,0,0,s_1,s_2}\rangle \ , \qquad (25)$$

$$\delta M_2^2 = \langle\phi_{1,0,0,s_3,s_4}|\frac{1}{2}\left[i\hat{Q},\hat{\mathcal{V}}^{(0)}\right]|\phi_{1,0,0,s_1,s_2}\rangle \ . \qquad (26)$$

First, δM_1^2:

$$\delta M_1^2 = \int d^3p\,d^3p'\langle\phi_{100}|\mathbf{p}'\rangle\langle\mathbf{p}|\phi_{100}\rangle\mathcal{V}^{(1)}(\mathbf{p}'s_3s_4;\mathbf{p}s_1s_2) \ . \qquad (27)$$

Using the rotational symmetry of the integrand, we can replace

$$\frac{(p_\perp - p'_\perp)^2}{(\mathbf{p}-\mathbf{p}')^2} \longrightarrow \frac{\frac{2}{3}\left((p_x - p'_x)^2 + (p_y - p'_y)^2 + (p_z - p'_z)^2\right)}{(\mathbf{p}-\mathbf{p}')^2} = \frac{2}{3} \ . \qquad (28)$$

After this, the remaining integrals are trivial (recall Eq. (9)) and we have

$$\frac{\delta M_1^2}{2m^2\alpha^4} = \frac{1}{2}\delta_{s_1s_2}\delta_{s_1s_4}\delta_{s_3s_4} - \frac{1}{12}\delta_{s_1\bar{s}_2}\delta_{s_3\bar{s}_4} + \frac{1}{2}\delta_{s_2\bar{s}_4}\delta_{s_1\bar{s}_2}\delta_{s_1\bar{s}_3} \ . \qquad (29)$$

Next, δM_2^2:

$$\delta M_2^2 = \langle\phi_{1,0,0,s_3,s_4}|\frac{1}{2}\left[i\hat{Q},\hat{\mathcal{V}}^{(0)}\right]|\phi_{1,0,0,s_1,s_2}\rangle \qquad (30)$$

$$= \frac{1}{2}\sum_{s_e s_{\bar{e}}}\int d^3p\,d^3p'\langle\phi_{100}|\mathbf{p}'\rangle\langle\mathbf{p}|\phi_{100}\rangle\left(\langle\mathbf{p}'s_3s_4|i\hat{R}|\mathbf{p}'s_e s_{\bar{e}}\rangle\langle\mathbf{p}'s_e s_{\bar{e}}|\hat{\mathcal{V}}^{(0)}|\mathbf{p}s_1s_2\rangle\right.$$

$$\left. -\langle\mathbf{p}'s_3s_4|\hat{\mathcal{V}}^{(0)}|\mathbf{p}s_e s_{\bar{e}}\rangle\langle\mathbf{p}s_e s_{\bar{e}}|i\hat{R}|\mathbf{p}s_1s_2\rangle\right) \ . \qquad (31)$$

Recalling Eq. (12) and using Eq. (23) we have

$$\delta M_2^2 = \frac{\alpha}{\pi^2}\int d^3p\,d^3p'\langle\phi_{100}|\mathbf{p}'\rangle\langle\mathbf{p}|\phi_{100}\rangle\frac{F}{(\mathbf{p}-\mathbf{p}')^2} \ , \qquad (32)$$

where

$$F = \sum_{s_e s_{\bar{e}}}\langle\mathbf{p}s_e s_{\bar{e}}|i\hat{R}|\mathbf{p}s_1s_2\rangle\langle\mathbf{p}'s_3s_4|\hat{v}^{(0)}|\mathbf{p}s_e s_{\bar{e}}\rangle \qquad (33)$$

$$= \frac{1}{2}\sum_{s_e s_{\bar{e}}}\left(\langle\mathbf{p}s_e s_{\bar{e}}|i\hat{R}|\mathbf{p}s_1s_2\rangle - \langle\mathbf{p}'s_e s_{\bar{e}}|i\hat{R}|\mathbf{p}'s_1s_2\rangle\right)\langle\mathbf{p}'s_3s_4|\hat{v}^{(0)}|\mathbf{p}s_e s_{\bar{e}}\rangle \ , \qquad (34)$$

using the fact that $\hat{v}^{(0)}$ is odd under $\mathbf{p}\longleftrightarrow\mathbf{p}'$ in this last step. Using Eq. (23) this becomes

$$F = \frac{1}{4m}\sum_{s_e s_{\bar{e}}}v^{(0)}(\mathbf{p}'s_3s_4;\mathbf{p}s_e s_{\bar{e}})v^{(0)}(\mathbf{p}'s_e s_{\bar{e}};\mathbf{p}s_1s_2) \ . \qquad (35)$$

Using the even symmetry of the rest of the integrand under the operations $(p_x \longrightarrow -p_x, p'_x \longrightarrow -p'_x)$ and $(p_x \longleftrightarrow p_y, p'_x \longleftrightarrow p'_y)$ this sum can be simplified with result

$$F = -\frac{1}{24m}(3g_1 + g_2)(\mathbf{p} - \mathbf{p}')^2, \tag{36}$$

where

$$g_1 = s_1 s_3 + s_2 s_4, \tag{37}$$

$$g_2 = 1 + s_1 s_2 - s_2 s_3 - s_1 s_4 + s_3 s_4 + s_1 s_2 s_3 s_4. \tag{38}$$

Recall that $s_i = \pm 1$, $(i = 1, 2, 3, 4)$; the '$\frac{1}{2}$' has been factored out of these spins.[5] The result was written in this form to show the equivalence with [1]. Combining the results we have

$$\delta M_2^2 = -\frac{\alpha}{24\pi^2 m}(3g_1 + g_2)\int d^3p\, d^3p' \langle \phi_{100}|\mathbf{p}'\rangle\langle\mathbf{p}|\phi_{100}\rangle \tag{39}$$

$$= -\frac{m^2\alpha^4}{24}(3g_1 + g_2), \tag{40}$$

using Eq. (9) in this last step.

Combining the results we have

$$\frac{\delta M_1^2 + \delta M_2^2}{2m^2\alpha^4} = \frac{1}{2}\delta_{s_1 s_2}\delta_{s_1 s_4}\delta_{s_3 s_4} - \frac{1}{12}\delta_{s_1 \bar{s}_2}\delta_{s_3 \bar{s}_4} + \frac{1}{2}\delta_{s_2 \bar{s}_4}\delta_{s_1 \bar{s}_2}\delta_{s_1 \bar{s}_3}$$

$$- \frac{1}{48}(3g_1 + g_2). \tag{41}$$

The eigenvalues are

$$\left\langle 1 \left| \delta M_1^2 + \delta M_2^2 \right| 1 \right\rangle = -\frac{5}{3}m^2\alpha^4, \tag{42}$$

$$\left\langle 2 \left| \delta M_1^2 + \delta M_2^2 \right| 2 \right\rangle = \frac{2}{3}m^2\alpha^4, \tag{43}$$

$$\left\langle 3 \left| \delta M_1^2 + \delta M_2^2 \right| 3 \right\rangle = \frac{2}{3}m^2\alpha^4, \tag{44}$$

$$\left\langle 4 \left| \delta M_1^2 + \delta M_2^2 \right| 4 \right\rangle = \frac{2}{3}m^2\alpha^4, \tag{45}$$

with corresponding eigenvectors

$$\left\{ |1\rangle = \frac{|+-\rangle - |-+\rangle}{\sqrt{2}}, \ |2\rangle = \frac{|+-\rangle + |-+\rangle}{\sqrt{2}}, \ |3\rangle = |--\rangle, \ |4\rangle = |++\rangle \right\}.$$

These results translate to the well known answer, $\frac{7}{6}\alpha^2 Ryd$, as can be seen by recalling the definition of our zeroth order and exact mass squared:

$$(2m + B_N)^2 = 4m^2 + 4mB_N + \delta^{(1)}M^2 + \mathcal{O}\left(m^2\alpha^5\right), \tag{46}$$

which gives

$$B_{triplet} - B_{singlet} = \frac{7}{6}\alpha^2 Ryd + \mathcal{O}\left(m\alpha^5\right), \tag{47}$$

the desired result.

---

[5]In order to get these simple forms for g_1 and g_2 it was useful to note the following simple relation: $\delta_{ss'} = \frac{1}{2}s(s + s')$ (true because $s^2 = 1$).

ACKNOWLEDGMENTS

The author wishes to thank the organizers of *Orbis Scientiae 1997* for providing such a stimulating atmosphere in which to work. The author would also like to thank Brent H. Allen, Martina M. Brisudová, Stanisław D. Głazek, Robert J. Perry, David G. Robertson and Kenneth G. Wilson for all the useful discussions that led to some simplifications in the calculations, and to a whole lot of understanding. In particular, the author would like to thank Brent H. Allen for the use of his matrix element rules [5]. Research reported in this paper has been supported by the National Science Foundation under grant PHY–9409042.

References

[1] B. D. Jones, R. J. Perry, and S. D. Głazek, 1996, hep-th/9605231, to appear in Phys. Rev. D.

[2] V. B. Berestetskii, E. M. Lifshitz and L. P. Pitaevskii, *Quantum Electrodynamics* (Pergamon Press, USA, Second edition 1982).

[3] M. Kaluza and H. J. Pirner, Phys. Rev. D **47**, 1620 (1993).

[4] M. M. Brisudová and R. J. Perry, Phys. Rev. D **54**, 6453 (1996).

[5] B. H. Allen, "Light-Front Time-Ordered Perturbation Theory Rules," OSU technical report, 1997 (unpublished).

[6] S. S. Schweber, *QED and the Men Who Made it: Dyson, Feynman, Schwinger, and Tomonaga* (Princeton University Press, New Jersey, 1994), p. 312.

SECTION VII

CURRENT EXPERIMENTS IN HIGH ENERGY PHYSICS

SEARCH FOR NEW PARTICLES WITH DELPHI AT LEP2

W. Adam for the DELPHI collaboration

Institut für Hochenergiephysik der ÖAW
Nikolsdorferg. 18, A-1050 Wien, Austria

INTRODUCTION

In late 1995 the physics programme at the LEP accelerator at CERN entered a new era with the first collisions at energies substantially above the Z^0 resonance. In 1996 LEP provided the first data sets at and above the W pair production threshold. These runs were used by the DELPHI collaboration to extend the searches for new particles beyond the sensitive region of LEP1.

A subject of primary interest in the analysis of these data sets is the search for the Higgs boson. In minimal extensions of the Standard Model (SM) two Higgs doublets are predicted, giving rise to a pair of charged Higgs bosons ($H^{\pm}$), two CP-even scalars (h and H), and a CP-odd pseudoscalar A. In a specific extension of this kind, namely the Minimal SuperSymmetric Model (MSSM), the associated production of hA is favoured at large values of $\tan \beta$, the ratio of the vacuum expectation values of the two doublets. In contrast to this, the cross section for pair production of charged Higgs bosons depends only on their mass. Results for these two channels will be presented.

Further analyses in the framework of supersymmetric theories concern the production of charginos, the fermionic partners of W bosons and the charged Higgs bosons mentioned above, and neutralinos, the supersymmetric partners of neutral gauge bosons and neutral Higgs states. Usually the lightest neutralino, $\tilde{\chi}_1^0$, is assumed to be the lightest supersymmetric particle (LSP). Under the assumption of R-parity conservation, which will be made throughout this paper, it represents the final state in the decay chain of supersymmetric particles. However, results on the production of charginos and neutralinos will also be presented in the scenario of a light gravitino acting as the LSP.

Finally results of searches for other kinds of "New Physics" will be presented in the form of an analysis aiming for the detection of single or double production of excited leptons. These states are assumed to be much heavier than ordinary leptons, and to decay promptly to their ground state via radiation of a γ, Z^0 or W.

THE EXPERIMENTAL APPARATUS

DELPHI is one of the four general purpose experiments operating at LEP. Design and performance of the detector have already been described elsewhere [1]. The components relevant for the analyses quoted below were

- the tracking system, consisting of a silicon microvertex detector, an inner detector in jet chamber structure, a time projection chamber and, further out from the vertex, additional drift chambers in the forward and barrel parts,

- the electromagnetic calorimeters in the forward and, with high spatial resolution, in the barrel part,

- the hadronic calorimeter, which is embedded in the return yoke of

- the superconducting coil providing a 1.2T magnetic field,

- the muon identification system, consisting of drift chambers within and outside the return yoke, and

- the luminosity monitors.

In preparation of the high energy runs of LEP an important upgrade programme has been undertaken. The acceptance for precise measurement of impact parameters and reconstruction of secondary vertices, and consequently for high efficiency b-tagging, has been enlarged by an extension of the vertex detector. The three cylindrical layers of this detector, mostly equipped with two-dimensional readout, allow for b-tagging down to an angle of 25° from the beam axis.

In order to improve redundancy, and thus efficiency, in the tracking for the very forward and backward regions conical endcaps have been added to the vertex detector, consisting of two layers of pixel and two layers of large pitch strip detectors, all of which provide measurement of impact points in two co-ordinates. Together they extend the sensitive region of the silicon tracker down to 10° from the beam pipe.

Finally the hermiticity of the detector, which is a key issue in many search channels, has been improved by adding scintillator counters in sensitive regions like the cable duct and the space between modules of the barrel electromagnetic calorimeter.

THE DATA SAMPLES

The first data above the Z^0 resonance were taken in late 1995 - about 6 pb^{-1} were recorded at centre-of-mass energies between 130 and 136 GeV. The results quoted in the following sections, however, refer mainly to data taken in 1996 at and above the WW threshold. About 10 pb^{-1} were recorded at an energy of 161 GeV, and a similar amount at 172 GeV.

Signal simulation samples for the Higgs searches were produced using the HZHA[2] and PYTHIA[3] generators. Chargino and neutralino samples were generated using SUSYGEN[4], partially in a modified version to include decays into a light gravitino. Excited leptons in single or double production were generated according to the cross sections in reference [5].

Standard model processes for background estimates using a variety of generators, in particular PYTHIA, with EXCALIBUR[6] for cross checks on four fermion processes and TWOGAM[7] for $\gamma\gamma$ processes. Quark fragmentation was modelled using JETSET 7.4[3], and all generated events were passed through the full detector simulation and analysis chain.

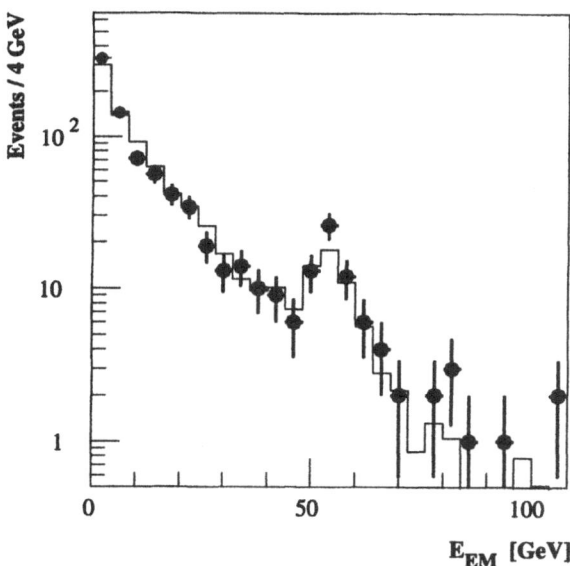

Figure 1. Electromagnetic energy of τ jet candidates in the $H^+H^- \rightarrow cs\tau\nu_\tau$ analysis in data (full points) and simulation (solid line), before rejection of ISR events. The clearly visible peak indicates the presence of ISR photons in the sample.

SEARCH FOR H^+H^-

In e^+e^- collisions charged Higgs bosons are mainly produced via s-channel Z^0 exchange. The cross section for this process is strongly dependent on the Higgs mass. In this analysis decays into cs and $\tau\nu_\tau$ were both searched for at a centre-of-mass energy of 161 GeV. In the following paragraphs the part aiming at the semileptonic final state will be discussed in some detail, since it illustrates the selection of a τ in an hadronic environment.

For the $cs\tau\nu_\tau$ state, hadronic events were preselected asking for a minimum charged multiplicity of 7, a charged (total) energy in the event of at least $15\%\sqrt{s}$ ($25\%\sqrt{s}$) and an acollinearity of at least $9°$. Several cuts were designed to reduce the important background from radiative events - due to the high cross section at the Z^0 resonance there is a high probability for a return to the Z^0 mass via initial state radiation (ISR).

After clusterisation into three jets the τ jet was assumed to be the one of lowest charged multiplicity. This choice was confirmed by further jet multiplicity and energy cuts. Again radiative events had to be rejected, since electromagnetic showers from an ISR photon can resemble a low multiplicity jet (figure 1). In order to reduce this background, photon conversion candidates were rejected, and cuts were applied to the energies of electromagnetic clusters associated to the τ jet candidate.

For optimum discrimination power, a linear discriminating function à la Fisher was then used, involving topological variables. On the events passing this filter a kinematical fit was applied, forcing four-momentum conservation and equal masses of the hadronic and the (τ + missing momentum) system. Events with poor fit quality were rejected, and a mass window around hypothetical Higgs masses was applied, designed to keep in each case 90% of the signal. The efficiency was determined from simulation to be 47–52%, at an expected background of 1.1–1.7 events. No candidate was found in data for masses up to 60 GeV/c^2.

In the purely hadronic final state, one candidate event was found at a mass of 46.2 GeV/c^2, compatible with background estimates. No event was selected for the purely

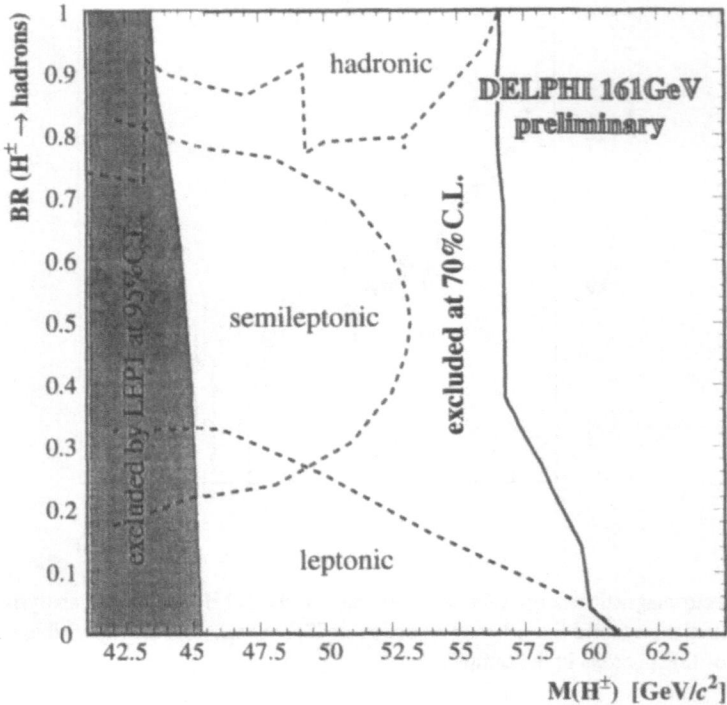

Figure 2. Exclusion regions for $M_{H\pm}$ as a function of BR(H$^\pm$ → hadrons). The dashed lines indicate the regions excluded at 70% confidence level by the search for hadronic (upper region), semileptonic (intermediate region) and leptonic (lower region) final states. The discontinuity in the upper region is due to the selected four jet event. The full line is obtained from the combination of the three results. The grey area indicates the 95% confidence level limit from LEP1.

leptonic channel. Combining the three results exclusion limits on $M_{H\pm}$ can be derived as a function of the hadronic branching ratio (figure 2). At 70% confidence level, charged Higgs bosons can be excluded up to masses of 56.4 GeV/c^2.

SEARCH FOR hA → b$\bar{\text{b}}$b$\bar{\text{b}}$

The CP-even and CP-odd neutral Higgs bosons, h and A respectively, are mainly produced together via e$^+$e$^-$ annihilation into Z^0 . For large values of tan β the cross section increases strongly with decreasing M_A and reaches values accessible with present data statistics. In this case the masses of the two bosons are nearly equal. Since the decay into b-quarks is dominating, the analysis aimed at the detection of the 4b final state. The branching ratio into this state amounts to about 85%.

Four jet events were selected using a hadronic preselection with rejection of radiative events, verification of a topology with at least four jets but without soft gluon jets, and a χ^2 cut on a kinematic fit imposing four-momentum conservation. After a loose b-tagging on the event level, the jet pairing with the lowest mass difference was chosen. A tight b-tagging on the jet level was then applied to reduce the background from q$\bar{\text{q}}$gg events. The selection was finally restricted to the high mass range, asking for a sum of the two dijet masses of at least 80 GeV/c^2.

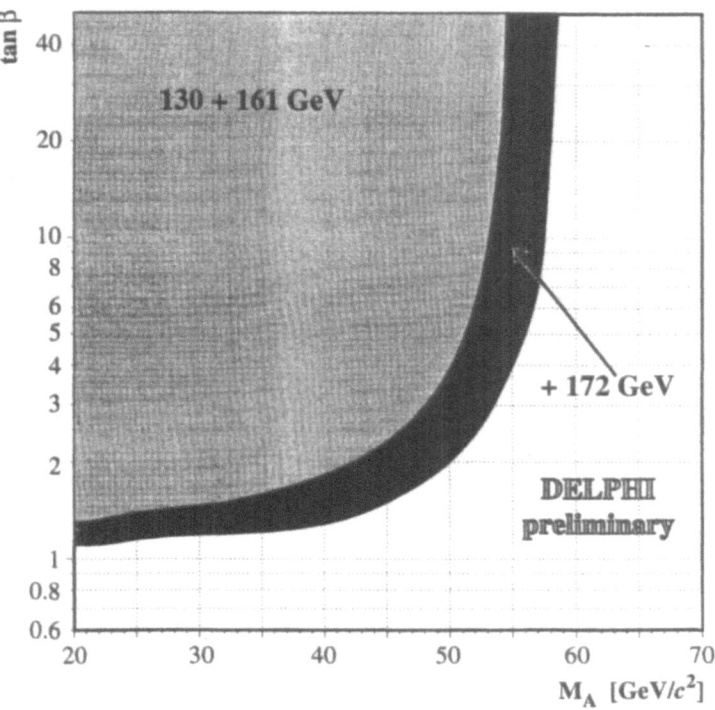

Figure 3. Excluded region in the the ($\tan \beta, M_A$) plane for the combination of 130–136 and 161 GeV results (light grey area), and for the preliminary combination with 172 GeV results (dark grey area) in the hA $\rightarrow$ b$\bar{\text{b}}$b$\bar{\text{b}}$ analysis.

No candidate was found in the 161 GeV sample, at an expected background of 0.6 events and an efficiency of 40.2%. Also in a preliminary analysis of 172 GeV data no event passed the cuts. The corresponding exclusion limits for M_A as a function of $\tan \beta$ are shown in figure 3.

SEARCH FOR CHARGINOS

Charginos, i.e. mixtures of the fermionic partners of W$^\pm$ and H$^\pm$, can be produced either via s-channel annihilation diagrams, or via t-channel exchange of a $\tilde{\nu}_e$. They decay into the LSP, which is here assumed to be the $\tilde{\chi}_1^0$, e.g. via emission of a W$^\pm$. The resulting topologies are of the $\ell\ell$, $jj\ell$ or multijet type, all with missing energy due to the LSPs escaping detection.

Event characteristics change significantly with the mass difference to the LSP ($\Delta M = M_{\tilde{\chi}^\pm} - M_{\tilde{\chi}^0}$) and the corresponding energy available for the visible decay products. Therefore the analysis was optimised for two different domains: $\Delta M \leq 10\text{GeV}/c^2$ ("degenerate") and $\Delta M > 10\text{GeV}/c^2$ ("non-degenerate").

Key variables of the selection were missing transverse momentum, which discriminates against $\gamma\gamma$ events, and acoplanarity. Events with high energy fractions in the forward / backward cones were rejected. In the leptonic topologies, an upper limit on the highest momentum charged particle was used against background from W$^+$W$^-$. The number of radiative events was further reduced by a veto on isolated, high energy photons, or activity in the hermeticity counters close to the missing momentum vector.

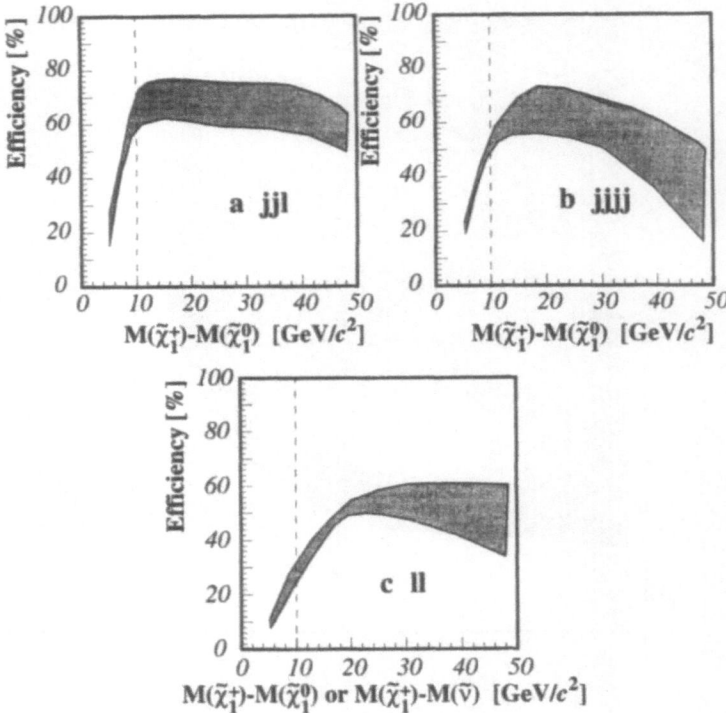

Figure 4. Efficiencies for the chargino search at 172 GeV for the semileptonic (a), hadronic (b) and leptonic (c) final states, as a function of the mass difference to the LSP. The bands indicate the spread due to the statistical error of the simulation and the dependence on the chargino mass. The dashed line indicates the limit between the "degenerate" (left) and "non-degenerate" (right) case.

At 172 GeV one multijet candidate was found for the degenerate case (expected background 1.1 ± 0.1 events), and a leptonic one for the non-degenerate case (expected background 3.3 ± 0.3 events). The efficiencies, as shown in figure 4, are mainly a function of ΔM.

These numbers can be converted into chargino mass limits (figure 5). In the degenerate case, the limit on the cross section corresponds to $M_{\tilde{\chi}^\pm} < 80 \text{GeV}/c^2$. In the non-degenerate case, destructive interference of the s- and t-channel contributions can occur, depending on the mass of the $\tilde{\nu}$ [8]. For low values of $M_{\tilde{\nu}_e}$ a limit of $M_{\tilde{\chi}^\pm} < 71.3 \text{GeV}/c^2$ can be derived.

SEARCH FOR NEUTRALINOS

This analysis aimed at the detection of $\tilde{\chi}^0_{i(i=2,3)}\tilde{\chi}^0_1$ production with subsequent $\tilde{\chi}^0_i \to \tilde{\chi}^0_1 \text{f} \bar{\text{f}}$ decay. The events are thus characterised by $\ell\ell$ or $\jmath\jmath$ topologies, both with missing energy.

For the hadronic case, two jet events were selected. No high energy, isolated charged particles were allowed, in order to lower background from W^+W^- production. Background from radiative and $\gamma\gamma$ events was reduced by a forward energy veto, selection of acoplanar events, and cuts on missing p^t, visible and missing mass. Finally a logical OR of three different selections was used to maintain high efficiency over a wide range of $\Delta M = M_{\tilde{\chi}^0_i} - M_{\tilde{\chi}^0_1}$ values.

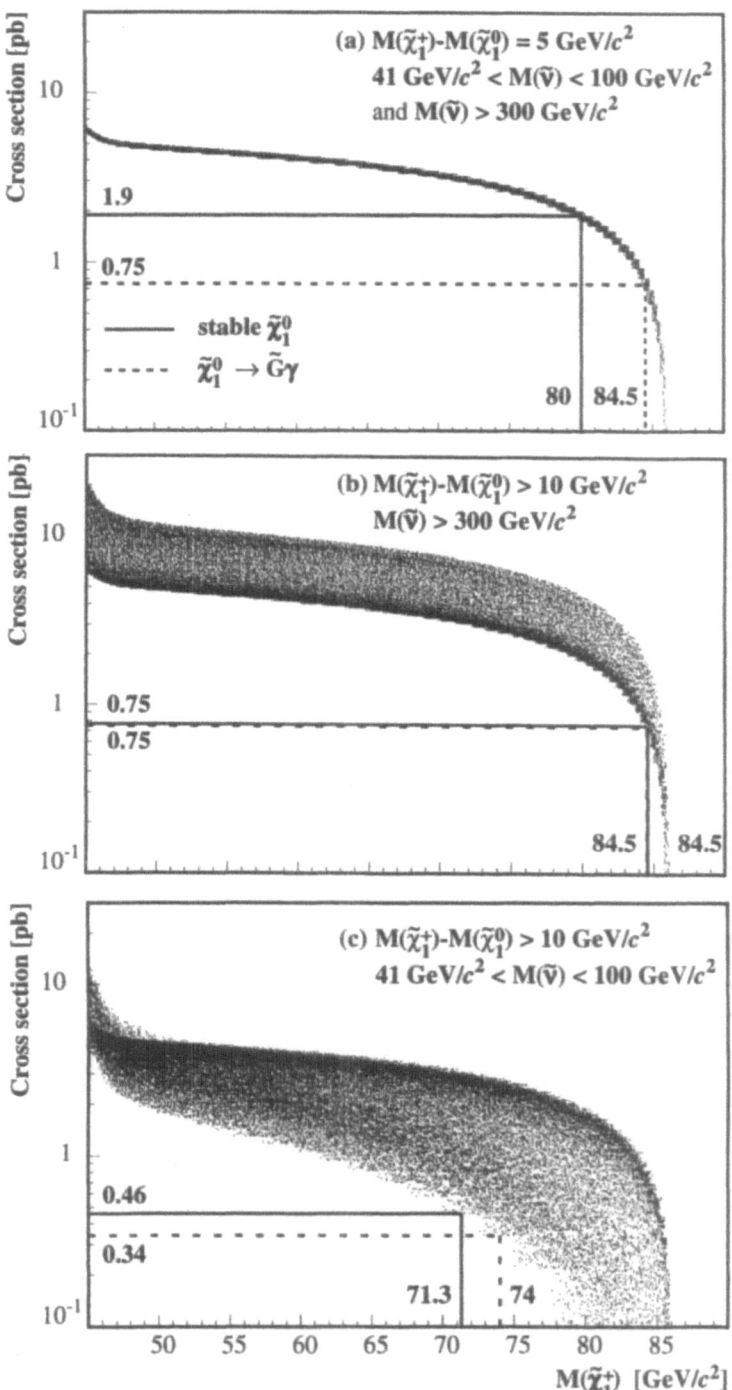

Figure 5. Preliminary chargino cross section and mass limits obtained from 161 and 172 GeV data. The points indicate the expected cross sections for different values of the parameters M_2, μ and $\tan\beta$. Figure (a) shows the case of small $\Delta M = M(\tilde{\chi}^+) - M(\tilde{\chi}^0)$, with small dependence of the cross section on the mass of the sneutrino. Figure (b) and (c) show the cross sections for $\Delta M > 10$ GeV/c^2 for high and low sneutrino masses. The full lines correspond to the $\tilde{\chi}_1^0$ being the LSP, the dashed ones to the light gravitino scenario (see chapter "Charginos and neutralinos in a light gravitino scenario").

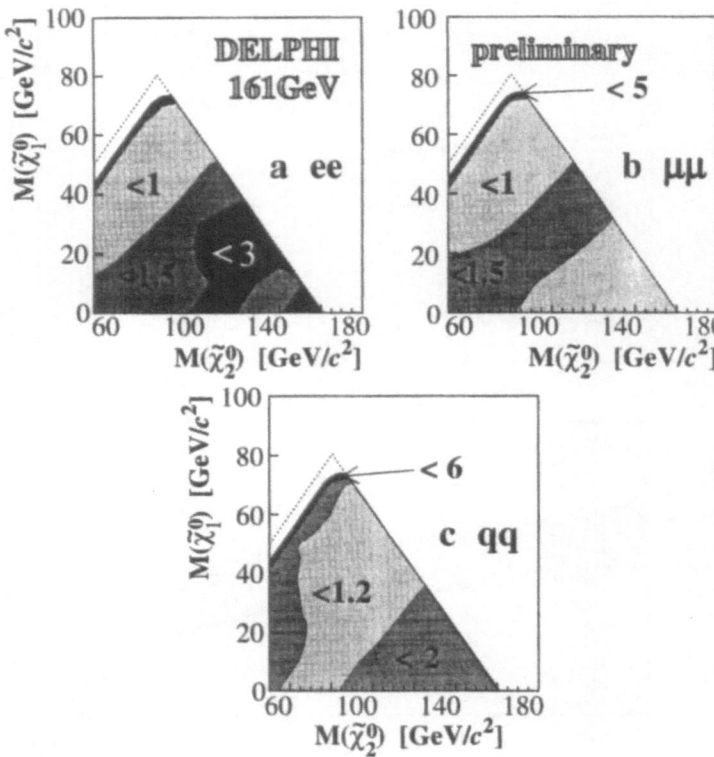

Figure 6. Preliminary cross section limits for $\tilde{\chi}_1^0 \tilde{\chi}_2^0$ production at 161 GeV. Figures (a-c) show 95% confidence level limits for the electron, muon and hadronic channels. The numbers associated with each area indicate the limits in pb.

For decays into leptons events with two isolated, oppositely charged particles were selected, at least one of which had to be loosely identified as an e or μ. The number of radiative and $\gamma\gamma$ events was reduced using a series of cuts on forward energy, multiplicity, acoplanarity, missing energy and missing p^t. A limit was imposed on the most energetic particle to reduce the W^+W^- background.

Two events were selected in the hadronic, and one in the leptonic topology, at estimated backgrounds of 2.6 ± 0.2 respectively 1.7 ± 0.7 events. Using the efficiencies, which were estimated using simulated samples with different mass combinations, cross section limits can be derived in the $M_{\tilde{\chi}_2^0}$–$M_{\tilde{\chi}_1^0}$ plane (figure 6). Exclusion zones in terms of the MSSM parameters μ and M_2 are obtained using a combination of neutralino and chargino results for centre-of-mass energies of 130–136 and 161 GeV (figure 7).

CHARGINOS AND NEUTRALINOS IN A LIGHT GRAVITINO SCENARIO

Recently scenarios involving a light gravitino have received renewed interest. In this case the gravitino acts as LSP, and $\tilde{\chi}^0$ decays into $\gamma\tilde{G}$ are possible[9]. For gravitino masses below about $250\,\mathrm{eV}/c^2$ these decays can occur within the detector.

In this scenario the chargino signatures as described above are complemented by two isolated photons. The selection was therefore extended by asking for the presence of two photons. Cuts against radiative events were released. The tagging of the two photons gave a better handle in the degenerate case and resulted in an efficiency of about 50%, down to low values of ΔM. No event was found in data, at an expected background of 1.1 events. The corresponding mass limits are shown, together with the results of the standard scenario, in figure 5.

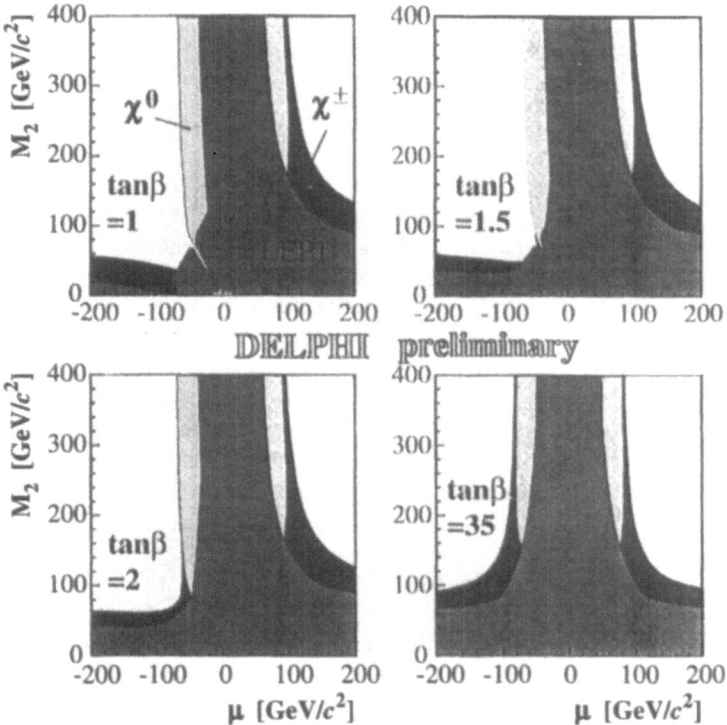

Figure 7. Regions in the (μ, M_2) plane excluded at 95% confidence level by chargino and neutralino search results at centre-of-mass energies of 130–136 and 161 GeV, for four different values of $\tan \beta$. The light shading shows the regions excluded by the neutralino search, while the dark shading corresponds to the combination of both results. The intermediate shading shows the areas excluded by LEP1 results.

In the neutralino case $\tilde{\chi}_1^0 \tilde{\chi}_1^0$ events become detectable using the two photons from the subsequent decays into $\tilde{G}$. In the standard model this topology can be produced in $e^+e^- \to \gamma\gamma(\gamma)$ and $e^+e^- \to \nu\bar{\nu}\gamma$ processes. Events with two photons of at least 10 GeV for the high energetic respectively 5 GeV for the low energetic photon were selected. At least one of the photons had to be detected in the barrel electromagnetic calorimeter. Events with charged tracks were vetoed, unless these tracks originated from the conversion of one of the photons. Acoplanarity, acollinearity and total energy are used to discriminate mainly against the QED background, the cross section of the $\nu\bar{\nu}\gamma$ process being small in the barrel region. The values of these cuts were tuned for a low and a high $M_{\tilde{\chi}^0}$ case. The sensitivity of the analysis

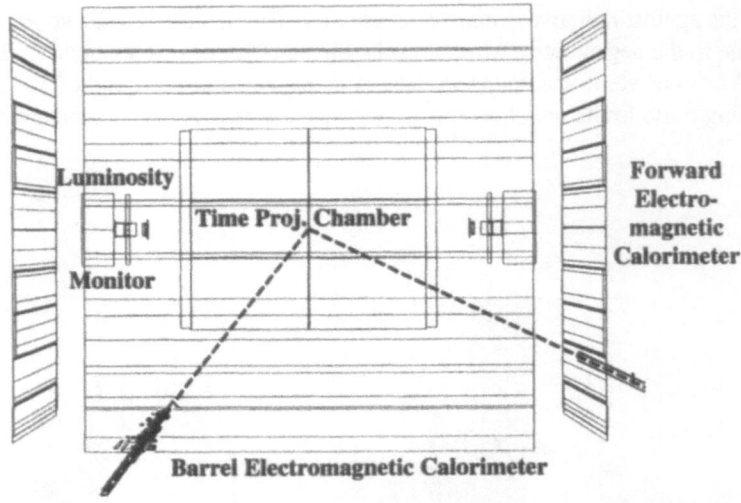

Figure 8. Example of an event with two acoplanar photons with energies of 39.2 and 29 GeV and a missing mass of 94.1 GeV/c^2. The dashed lines indicate the direction from the vertex to the showers in the barrel respectively forward electromagnetic calorimeters.

was limited to decay lengthes of up to one metre. At efficiencies of 20–50% a total of three candidates were found in the 130–172 GeV samples, at an estimated background of 2.1 (2.3) events for the low (high) $M_{\tilde{\chi}^0}$ case. An example of such an event is shown in figure 8.

SEARCH FOR EXCITED LEPTONS

In e^+e^- collisions both single and double production of excited leptons are possible. In the corresponding Lagrangian[5] the standard model lepton couplings are modified by a factor f for the charged and f' for the neutral current. Single production occurs via s-channel Z^0 or γ exchange - in the case of e^* and ν_e^* additional t-channel diagrams are involved. Double production occurs dominantly via s-channel diagrams. Excited leptons decay into leptons via γ, Z^0 or W emission, resulting in a multitude of different final states. The branching ratios depend strongly on the values, and in particular the relative sign, of f and f'.

The final states covered in this analysis were $\gamma\gamma\nu\nu$, $\ell\ell\nu\nu$, $\ell\ell\gamma$, $\ell\ell\gamma\gamma$, $\ell\ell\ell\ell$, $\ell\nu jj$, $\ell\ell jj$, $\ell\ell\ell\nu jj$ and $\ell\nu\nu\nu jj$ as well as single and double photon events.

In the common preselection events from $\gamma\gamma$ processes were removed. The sample was then divided according to charged multiplicity into "hadronic" ($N_{ch} > 8$) and "leptonic" ($N_{ch} \leq 8$) events. They were further classified using the number of jets with at least one charged particle, of isolated charged particles and of isolated photons. In this way the large variety of final states could be covered. Efficiencies ranged from 15% in the $\ell^* \to \nu$W case to more than 60% in $\ell^* \to \ell\gamma$ decays.

Results for centre-of-mass energies of 130 and 161 GeV have already been published[10]. Preliminary results are available from the combination with 172 GeV data. For the case of double production, mass limits have been set for two assumptions on the relative sign of the couplings and are summarised in table 1.

For the case of single production, limits on the ratio of coupling and mass are available as a function of the excited lepton mass (figure 9).

Table 1. Preliminary limits from 172 GeV data on the masses of excited charged leptons from the double production mode, under two assumptions on the relative couplings. Both left- and right-handed components have been assumed.

	e^*	μ^*	τ^*
$f = f'$	84.6 GeV/c^2	84.9 GeV/c^2	84.6 GeV/c^2
$f = -f'$	78.1 GeV/c^2	78.1 GeV/c^2	78.1 GeV/c^2

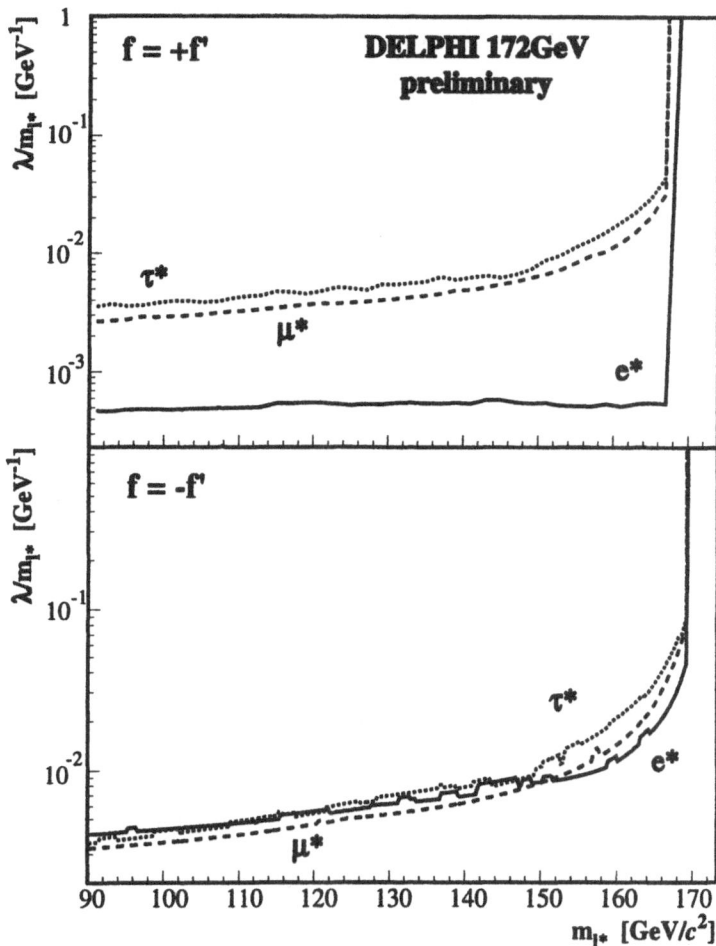

Figure 9. Limits at 95% confidence level on the ratio of coupling and mass of excited electrons (solid line), muons (dashed line) and taus (dotted line) for single production at 172 GeV.

CONCLUSIONS

A set of preliminary results is available from the analyses of data taken with the DELPHI detector at centre-of-mass energies up to 172 GeV, extending exclusion zones past the limits set by the LEP1 analyses. A new mass limit of 56.4 GeV/c^2, at 70% confidence level, has been obtained for charged Higgs bosons. The negative result of the search for hA $\rightarrow$ b$\bar{\text{b}}$b$\bar{\text{b}}$ has been interpreted in terms of an exclusion region in the ($\tan \beta, M_A$) plane.

Further results on production of supersymmetric particles have been presented. In the standard scenario, assuming $\tilde{\chi}_1^0$ as the LSP, cross section limits have been obtained for $\tilde{\chi}^+ \tilde{\chi}^-$ and $\tilde{\chi}_1^0 \tilde{\chi}_2^0$. In the non-degenerate case, and for low sneutrino masses, a 95% confidence level limit of 71.3 GeV/$c^2$is found for $M_{\tilde{\chi}_1^+}$. The results of both searches have been used to obtain exclusion regions in the (μ, M_2) plane. Furthermore no evidence was found for chargino or neutralino production in scenarios involving a light gravitino as the LSP.

Besides searches in the context of supersymmetric models results on single and double production of excited leptons have been presented, and limits on masses and couplings have been calculated.

Search for other kinds of "New Physics" in the high energy data of DELPHI is in progress. A further increase in sensitivity is expected from the combination of the results of the four LEP experiments by working groups on searches for Higgs bosons and supersymmetric particles. LEP operation will resume in 1997 with an increase in E_{CMS} of more than 10 GeV and will thus open the possibility to explore regions of even higher mass.

REFERENCES

1. DELPHI Coll., P.Abreu et al., *Nucl. Instr. Methods* A378:57 (1996).
2. P.Janot, in: *Physics at LEP2*, eds. G.Altarelli, T.Sjöstrand and F.Zwirner, CERN 96-01 vol.2, Geneva (1996).
3. T.Sjöstrand, *Comp. Phys. Comm.* 39:347 (1986).
4. S.Katsanevas and S.Melachroinos, in: Physics at LEP2, eds. G.Altarelli, T.Sjöstrand and F.Zwirner, CERN 96-01 vol.2, Geneva (1996).
5. F.Boudjema, A.Djouadi and J.L.Kneur, *Z. Phys.* C57:425 (1993).
6. F.A.Berends, R.Pittau, R.Kleiss, *Comp. Phys. Comm.* 85:437 (1995).
7. S.Nova, A.Olchevski and T.Todorov, DELPHI Note 90-35 PROG 152 (1990).
8. A.Bartl, H.Fraas, W.Majerotto and B.Mösslacher, *Z. Phys.* C55:257 (1992) and references therein.
9. D.Dicus et al., *Phys. Rev.* D43:2951 (1991).
   S.Ambrosiano et al., *Phys. Rev.* D54:5395 (1996) and references therein.
10. DELPHI Coll., P.Abreu et al., Phys. Lett. B380:480 (1996).
    DELPHI Coll., P.Abreu et al., CERN-PPE/96-169, to be published in *Phys. Lett.* B.

W PHYSICS RESULTS FROM DELPHI

Hywel T. Phillips[1], on behalf of the DELPHI collaboration

[1]Rutherford Appleton Laboratory
Chilton, Didcot, Oxfordshire
OX11 0QX. United Kingdom

INTRODUCTION

This note presents recent results obtained using the DELPHI detector at the Large Electron-Positron collider (LEP) at CERN in Geneva. The LEP machine, which ran at the Z peak from 1989 to 1995, was upgraded with superconducting radio-frequency cavities in 1995-6 in order to increase the available beam energy.

In 1996, LEP ran briefly at the Z for calibration purposes and then moved on to higher energies. Approximately 10 pb^{-1} were delivered to each of the four LEP experiments (ALEPH, DELPHI, L3 and OPAL) at centre-of-mass energies of 161.3 GeV and 172 GeV. 161.3 GeV is just above WW pair production threshold, which allowed the reaction $e^+e^- \rightarrow W^+W^-$ to be studied for the first time.

W Physics At LEP2: Motivation

The mass of the W boson is a fundamental parameter of the Standard Model. It has been measured directly in $p\bar{p}$ collisions[1] to be

$$M_W = 80.35 \pm 0.13 \text{ GeV/c}^2.$$

A Standard Model fit to data obtained at the Z peak during LEP1 gives

$$M_W = 80.352 \pm 0.033 \text{ GeV/c}^2.$$

Thus, a precise measurement of M_W provides a sensitive test of the self-consistency of the Standard Model. Furthermore, if the mass of the top quark is known, a precise measurement of M_W can give a handle on the mass of the Higgs boson[2].

The presence of diagrams containing the coupling of three gauge bosons at tree-level (shown in figure 1) allows sensitive tests of the Standard Model predictions for the couplings WWγ and WWZ to be performed at LEP2. The exact form of the non-

Abelian self-couplings of the W,Z and γ is one of the most crucial consequences of the SU(2)×U(1) theory of Electroweak interactions. The best limits on non-standard contributions to these trilinear gauge couplings (TGC's) to date have been set in $p\bar{p}$ experiments [3],[4],[5],[6]. However, studies during the LEP2 workshop indicated that LEP2 should ultimately be more sensitive[7].

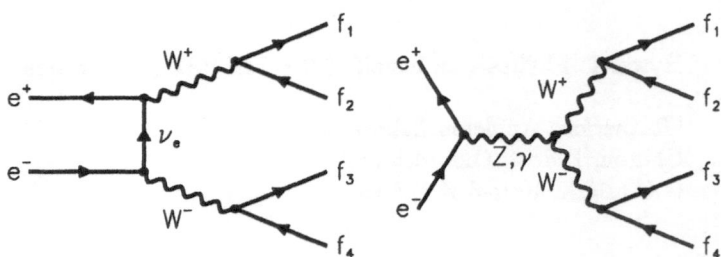

Figure 1. The tree-level graphs for WW production at LEP2.

W PHYSICS AT 161.3 GeV

The analysis summarised here is described in more detail in [8].

The cross-section for WW production just above threshold is very sensitive to M_W, and since M_W has been reasonably well-measured in $p\bar{p}$ experiments, an energy scan was not required. Instead, a single running point of $\sqrt{s} \sim 2M_W + 0.5$ GeV was found to be suitable[2] and hence the choice of an energy just above 161 GeV.

Within the framework of the Standard Model, a measurement of the threshold cross-section can be interpreted in terms of the W mass. Most of the variation in cross-section with M_W comes from the opening up of phase space – there is a relatively small dependence on the detailed dynamics of the process. This has unfortunate consequences for the sensitivity of TGC measurements made near threshold. In addition, the s-channel amplitudes containing all but one of the possible TGC's WW(γ,Z) are suppressed with respect to the t-channel neutrino exchange amplitude near threshold by a relativistic β factor.

W^+ bosons can decay to $e^+ \nu_e$, $\mu^+ \nu_\mu$, $\tau^+ \nu_\tau$, $u \bar{d}$, $c \bar{s}$, plus other quark combinations via CKM mixing. W^- decays are the charge conjugates of the above. Since there are two W bosons per event, there are three event types, examples of which are shown in figures 2, 3 and 4. In these figures, the beams go into and out of the paper. The detector is shown in outline form - proceeding out from the interaction point are the time projection chamber, outer tracking detector, electromagnetic calorimeter, hadronic calorimeter and muon chambers. Single curves represent the paths of charged particles within the detector, the curvature being induced by the DELPHI solenoidal magnetic field. Calorimeter depositions are shown as squares whose area is proportional to the energy deposited. The event types are:

1. Hadronic, where both W's decay to quarks. This has a predicted branching fraction of 46% in the Standard Model. An example of this sort of event is shown in figure 2.

2. Semileptonic, where one W decays to lepton-neutrino and the other decays to quarks. This has a predicted branching fraction of 44% in the Standard Model. An example of this sort of event is shown in figure 3.

3. Leptonic, where both W's decay to lepton-neutrino. This has a predicted branching fraction of 10% in the Standard Model. An example of this sort of event is shown in figure 4.

Backgrounds

The main background for WW studies is the reaction $e^+e^- \rightarrow q\bar{q} + n\gamma$. If an initial state photon of sufficient energy is emitted, the Z boson exchanged in the s-channel can be nearly on-shell. This gives a resonant enhancement to this part of the cross-section, a phenomenon called radiative return to the Z.

Other backgrounds include two-photon processes, which have large cross-sections but are predominantly soft and hence easy to reject. Other four-fermion neutral current processes e.g. ZZ production are a background to WW production. In some cases, these other processes can produce the same final state as WW production and thus must be included at the amplitude level in order to incorporate interference effects. Four-fermion generators such as ERATO[9] and EXCALIBUR[10] were used for this purpose.

Hadronic Channel

The following points give an outline of how hadronic WW events were selected:

- At least four jets were required, each containing at least three charged particles.

- The effective centre-of-mass energy $\sqrt{s'}$ was required to be at least 115 GeV, i.e. inconsistent with being a radiative return event. $\sqrt{s'}$ was reconstructed by assuming that each event consisted of two jets plus a photon: if an isolated photon was detected, its energy was used to reconstruct $\sqrt{s'}$, otherwise, a photon was assumed to have been lost down the beam pipe and the jet angles used to calculate $\sqrt{s'}$.

- After forcing the event to four jets, a kinematic fit was performed imposing four-momentum conservation. A discriminant variable was then used to remove gluon bremsstrahlung $q\bar{q}\gamma$ events.

A selection efficiency of $61.3 \pm 2.0\%$ was estimated from simulated events and the residual background cross-section was estimated to be 0.61 ± 0.07 pb. The dominant background remaining was from $q\bar{q}\gamma$ gluon bremsstrahlung events giving four or more jets.

A total of 15 events were selected and a fit to the discriminant variable including the expected background gave a cross-section in the hadronic channel, σ_{WW}^{4jet}, of

$$\sigma_{WW}^{4jet} = \sigma_{WW}^{total} \times BR(WW \rightarrow 4jets) = 1.56^{+0.67}_{-0.55} \pm 0.13 \text{ pb,}$$

where σ_{WW}^{total} is the total WW cross-section and $BR(WW \rightarrow 4jets)$ is the hadronic WW branching ratio. The first error quoted is statistical, the second systematic.

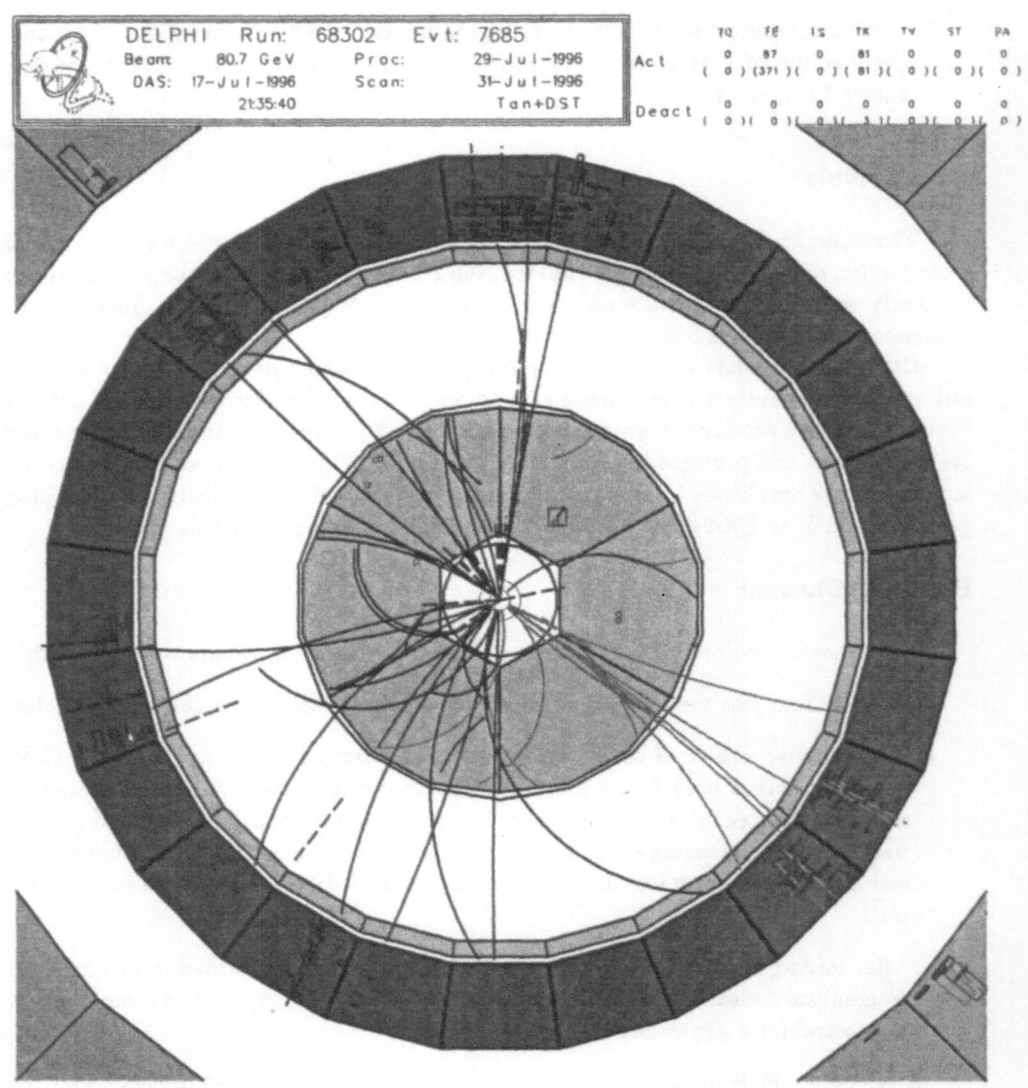

	TQ	FE	IS	TK	TV	ST	PA
Act	0	87	0	81	0	0	0
	(0)	(371)	(0)	(81)	(0)	(0)	(0)
Deact	0	0	0	0	0	0	0
	(0)	(0)	(0)	(3)	(0)	(0)	(0)

DELPHI   Run: 68302   Evt: 7685
Beam   80.7 GeV   Proc: 29–Jul–1996
DAS: 17–Jul–1996   Scan: 31–Jul–1996
           21:35:40   Tan+DST

Figure 2. An example of a hadronic WW event seen in the DELPHI detector. Four jets of particles are visible.

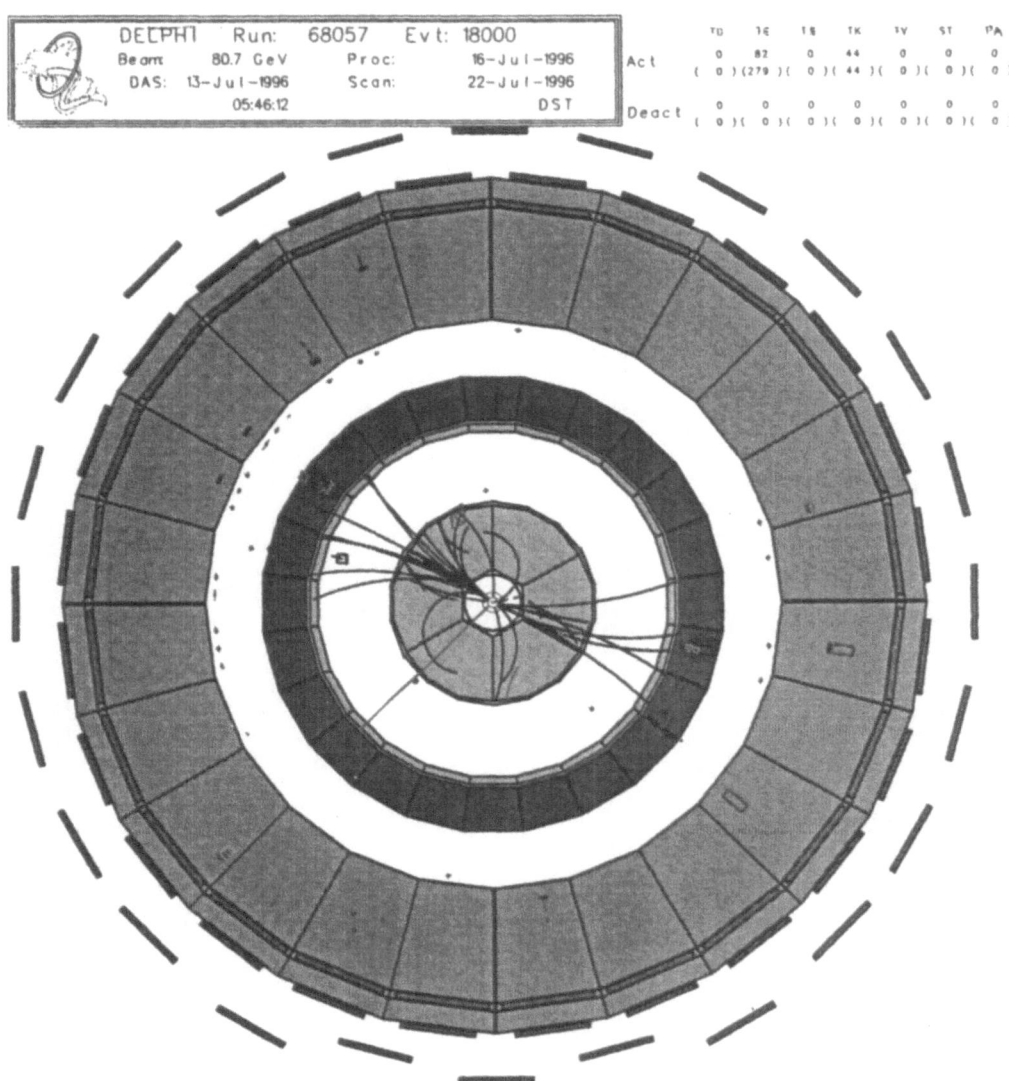

Figure 3. An example of a muon semileptonic WW event seen in the DELPHI detector. An isolated, stiff particle track is seen emerging from the interaction point and passing through all layers of the detector – this is a muon from W decay. Two back-to-back jets of particles come from the decay of the other W. The neutrino partner of the muon escapes detection, leaving a momentum imbalance in the detector.

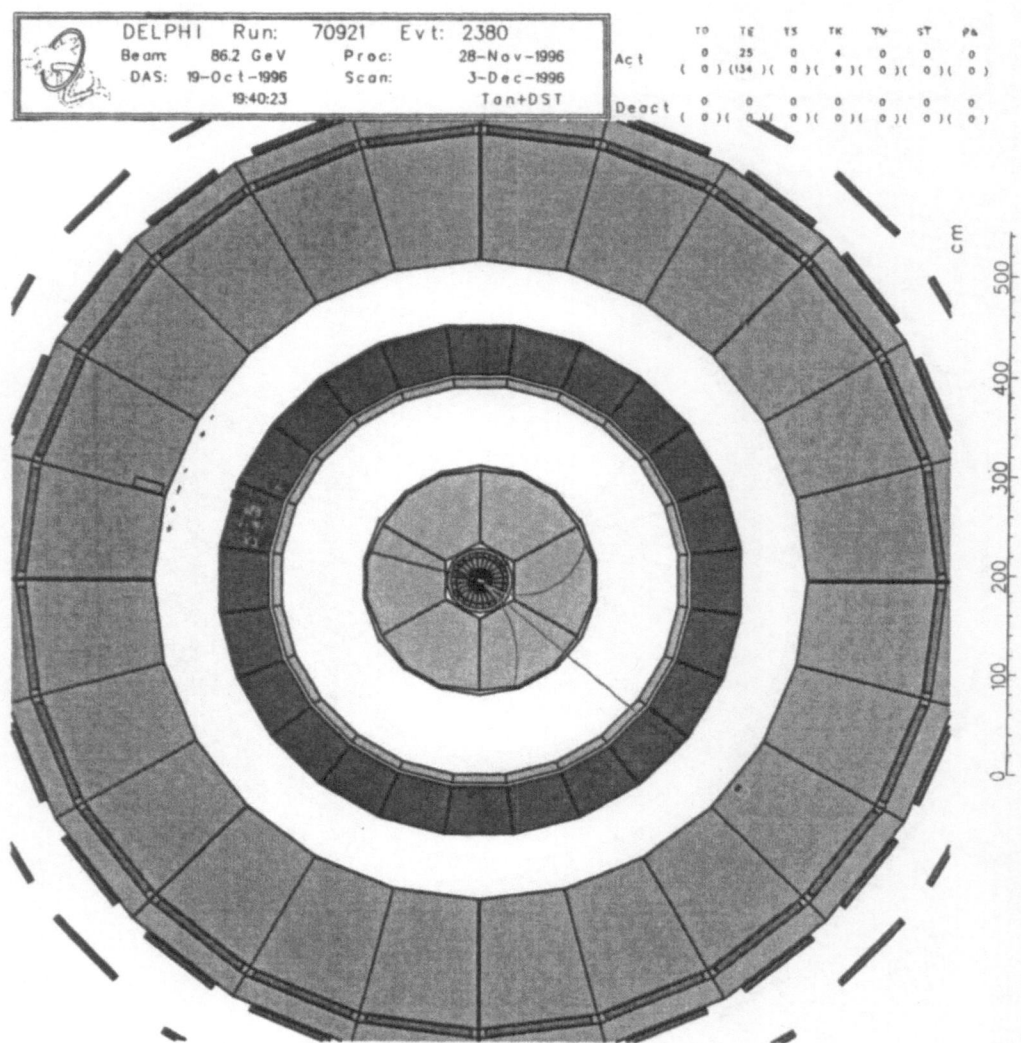

Figure 4. An example of a leptonic WW event seen in the DELPHI detector. One W has decayed to an electron, which is seen as a stiff track pointing to a large cluster of energy deposition in the electromagnetic calorimeter, plus a neutrino which escapes detection. The other W has decayed to a muon, seen as a stiff track emerging from the detector, plus an undetected neutrino. The two curved tracks near the muon are an electron-positron pair from a photon which converted in the beampipe.

Semileptonic Channel

The following points give an outline of how semileptonic WW events were selected:

- At least five charged particles were required to be seen in the detector.

- One and only one high-energy, isolated, tagged electron or muon was found. For tau events, an untagged selection with harder cuts was performed, searching for three-jet events with one narrow, tau-like jet.

- The missing momentum vector was required to point away from the beampipe and the cores of the jets and have a magnitude of at least 10 GeV/c.

- The hadronic part of the event was then forced to two jets; a jet-jet mass greater than 30 GeV/c² and jet-jet angle greater than 80° was required.

- Events with isolated clusters of energy in the electromagnetic calorimeters were rejected.

The selection efficiency was estimated to be $(60.9 \pm 3.0)\%$ from fully-simulated events and the residual background cross-section was estimated to be 0.193 ± 0.024 pb. A total of 12 events were selected, giving a cross-section in the semileptonic channel, $\sigma_{WW}^{l\nu jj}$, of

$$\sigma_{WW}^{l\nu jj} = \sigma_{WW}^{total} \times BR(WW \to l\nu jj) = 1.77^{+0.67}_{-0.55} \pm 0.10 \text{ pb},$$

where σ_{WW}^{total} is the total WW cross-section and $BR(WW \to l\nu jj)$ is the semileptonic WW branching ratio. The first error quoted is statistical, the second systematic.

Leptonic Channel

The following points give an outline of how leptonic WW events were selected:

- Between two and six charged particles were required to be seen in the detector and the scalar sum of their energies was required to be greater than 40 GeV.

- The event was then clustered into jets; subsequent selections used jet variables in order to include τ decays. Two well-separated high-energy jets were required and the event was required be both acollinear and acoplanar.

- The missing momentum vector was required to point away from the beampipe.

- If any isolated cluster of energy in the electromagnetic calorimeters exceeded 30 GeV, the event was rejected.

The selection efficiency was estimated to be $(47.7 \pm 3.0)\%$ from fully-simulated events (although substantially higher for events without τ) and the residual background cross-section was estimated to be 0.06 ± 0.04 pb. A total of 2 events were selected, giving a cross-section in the leptonic channel, $\sigma_{WW}^{l\nu l\nu}$, of

$$\sigma_{WW}^{l\nu l\nu} = \sigma_{WW}^{total} \times BR(WW \to l\nu l\nu) = 0.31^{+0.39}_{-0.24} \pm 0.09 \text{ pb},$$

where σ_{WW}^{total} is the total WW cross-section and $BR(WW \to l\nu l\nu)$ is the leptonic WW branching ratio. The first error quoted is statistical, the second systematic.

WW Cross-Section and W Mass

The results from the three different channels were combined by performing a likelihood fit to the product of the Poisson probabilities of seeing the observed number of events per channel as a function of M_W, constraining to the Standard Model WW branching fractions. This gave a total cross-section of

$$\sigma_{WW}^{total} = 3.67^{+0.97}_{-0.85} \text{ (stat.)} \pm 0.19 \text{ (syst.) pb.}$$

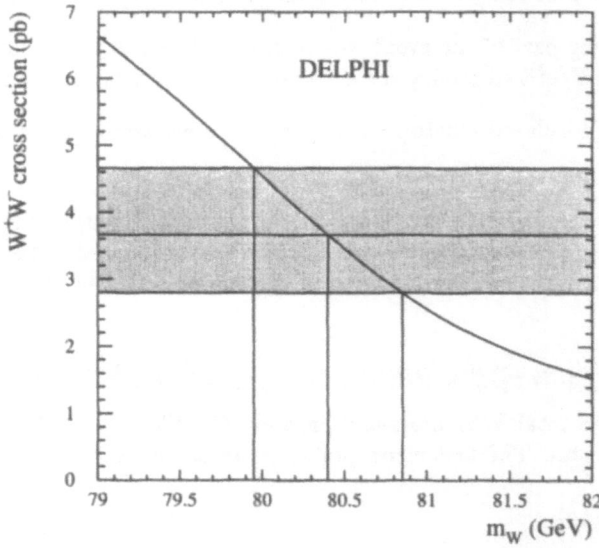

Figure 5. Standard Model prediction of the WW pair cross-section as a function of the W mass. The curve was generated using the GENTLE program. The horizontal line shows the observed cross-section and the shaded band its one standard deviation error. Reprinted from P.Abreu et al., Physics Letters B (1997) with kind permission from Elsevier Science - NL, Sara Burgerhartstraat 25, 1055 KV Amsterdam, The Netherlands.

Comparing this observed value with a theoretical curve of σ_{WW}^{total} vs. M_W, as shown in figure 5, gave a value for the W mass of

$$M_W = 80.40 \pm 0.44 \ (stat.) \pm 0.09 \ (syst.) \pm 0.03 \ (\text{LEP}) \ \text{GeV}/c^2,$$

where the first error is statistical, the second systematic and the third is the error due to the current uncertainty in the LEP beam energy.

The theoretical curve was obtained using the GENTLE program[11]. This program includes only the three tree-level WW production graphs of figure 1. The effects of additional graphs which produce the same final states as WW production (for example, the single W production graphs shown in figure 6) were taken into account by applying correction factors (of order 1) to the WW cross-sections. These correction factors were calculated using the EXCALIBUR Monte Carlo generator[10] and a simulation of the DELPHI detector. The figures previously quoted include these corrections.

PRELIMINARY COMBINED LEP RESULTS AT 161.3 GeV

The four LEP collaborations agreed to combine their observed WW production cross-section measurements and convert this into a measured value of the W mass. The following results are from the LEP Electroweak working group and are preliminary[12]. The resulting total cross-section was

$$\sigma_{WW}^{total} = 3.65 \pm 0.45 \text{ pb (LEP preliminary)}$$

giving

$$M_W = 80.42^{+0.22}_{-0.21} \text{ GeV}/c^2 \text{ (LEP preliminary)},$$

where the quoted error includes statistical and systematic errors added in quadrature. This is compatible with the values measured in other experiments quoted in the introduction.

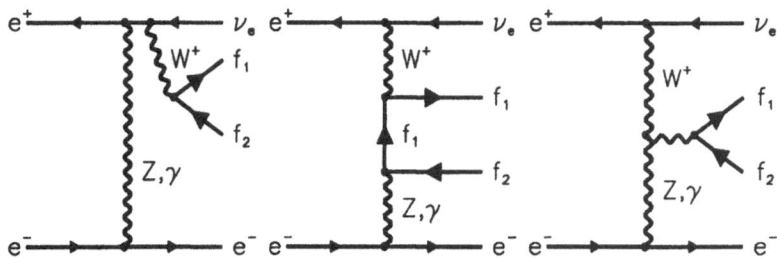

Figure 6. Examples of single W graphs at LEP2.

W PHYSICS AT 172 GeV

Using a set of selections similar to those used at 161 GeV (scaled with energy where appropriate), a preliminary value of the WW cross-section at 172 GeV of

$$\sigma_{WW}^{total} = 11.38^{+1.54}_{-1.43} \text{ (stat.)} \pm 0.32 \text{ (syst.) pb (DELPHI preliminary)}$$

was obtained. The variation of cross-section with centre-of-mass energy is shown in figure 7: good agreement was found between data and the predictions of the Standard Model.

The total cross-section at 172 GeV is not very sensitive to M_W. Instead, M_W can be measured by direct reconstruction, using the momenta of particles measured in the detector to reconstruct the mass of the W bosons in each event. This analysis is not yet complete, but the expected precision from the ~ 10 pb^{-1} of data taken by each experiment is similar to that obtained from the threshold cross-section measurement

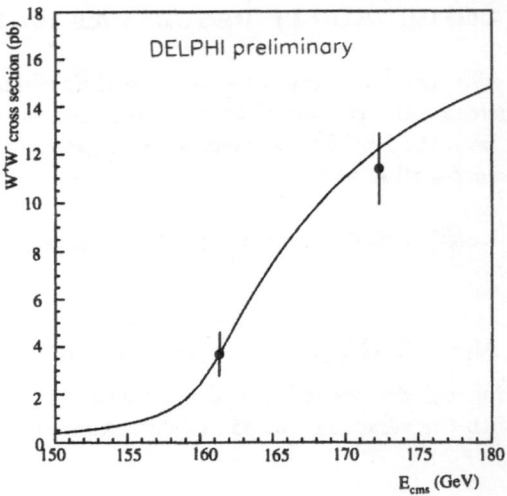

Figure 7. W pair cross-section as a function of centre-of-mass energy. The points with errors are data; the curve is the Standard Model prediction for $M_W = 80.36$ GeV/c^2.

at 161.3 GeV as studies during the LEP2 workshop showed that the two methods have comparable precision per inverse picobarn[2].

TRILINEAR GAUGE COUPLINGS

Model

The most general effective Lagrangian for the interaction WWV (V=γ,Z) is

$$iL_{effective}^{WWV} = g_{WWV} \left[\begin{array}{l} g_V^1 V^\mu \left(W_{\mu\nu}^- W^{+\nu} - W_{\mu\nu}^+ W^{-\nu} \right) + \\ \kappa_V W_\mu^+ W_\nu^- V^{\mu\nu} + \\ \frac{\lambda_V}{M_W^2} V^{\mu\nu} W_\nu^{+\rho} W_{\rho\mu}^- + \\ ig_V^5 \epsilon_{\mu\nu\rho\sigma} \left((\partial^\rho W^{-\mu}) W^{+\nu} - W^{-\mu} (\partial^\rho W^{+\nu}) \right) V^\sigma + \\ ig_V^4 W_\mu^- W_\nu^+ \left(\partial^\mu V^\nu + \partial^\nu V^\mu \right) - \\ \frac{\tilde{\kappa}_V}{2} W_\mu^- W_\nu^+ \epsilon^{\mu\nu\rho\sigma} V_{\rho\sigma} - \\ \frac{\tilde{\lambda}_V}{2M_W} W_{\rho\mu}^- W_\nu^{+\mu} \epsilon^{\nu\rho\alpha\beta} V_{\alpha\beta} \end{array} \right]$$

The couplings $g_V^1, \kappa_V, \lambda_V$ are CP conserving, the coupling g_V^5 is C and P violating but CP conserving and the couplings $g_V^4, \tilde{\kappa}_V, \tilde{\lambda}_V$ are CP violating.

In the Standard Model,

$$g_Z^1 = g_\gamma^1 = \kappa_Z = \kappa_\gamma = 1$$

and all other couplings are zero.

As there are so many possible couplings, a model of the new physics responsible for deviations from the standard couplings was adopted to reduce the total number of independent variables. The model was that recommended during the LEP2 workshop, described in detail in [7]. This leads to three independent CP conserving couplings

$$
\begin{vmatrix} g^1_\gamma \\ \kappa_\gamma \\ \lambda_\gamma \\ \kappa_Z \\ \lambda Z \\ g^5_Z \end{vmatrix} \xrightarrow{\text{dimension} \le 6, SU(2) \times U(1)} \begin{vmatrix} \alpha_{W\phi} \\ \alpha_W \\ \alpha_{B\phi} \end{vmatrix}
$$

and two independent CP violating couplings

$$
\begin{vmatrix} \tilde{\kappa}_\gamma \\ \tilde{\lambda}_\gamma \\ \tilde{\kappa}_Z \\ \tilde{\lambda}_Z \\ g^4_\gamma \\ g^4_Z \end{vmatrix} \xrightarrow{\text{dimension} \le 6, SU(2) \times U(1)} \begin{vmatrix} \tilde{\alpha}_{W\phi} + \tilde{\alpha}_{B\phi} \\ \tilde{\alpha}_W \end{vmatrix}
$$

which are $SU(2) \times U(1)$ invariant, arise from operators of dimension less than or equal to 6 and produce no large effects which would have been observed at LEP1. We have interpreted our results in terms of just two of these independent couplings: $\alpha_{W\phi}$, because this is the CP conserving coupling to which our sensitivity is greatest, and $\tilde{\alpha}_W$ because this coupling does not have the usual kinematic suppression factor at threshold [13].

Limits on TGC's from 161.3 GeV data

The Standard Model gives precise predictions of the WW cross-section near threshold. In previous sections, this was used as a framework for measuring the W mass. However, one can use another measurement of M_W to predict the cross-section at 161.3 GeV. The DELPHI cross-section measurement can then be interpreted in terms of limits on the contributions from non-standard TGC's.

We consider only one-parameter cases for $\alpha_{W\phi}$ and $\tilde{\alpha}_W$ with all other TGC's set to zero. As the amplitudes depend linearly on any TGC α_i, the cross-section has a quadratic dependence on the TGC. The minima of these quadratics occur near the standard model value of $\alpha_i = 0$. The predicted variation of the number of observed events was evaluated using the EXCALIBUR and ERATO Monte Carlo generators and a full simulation of the DELPHI detector. The expected number of events as a function of $\alpha_{W\phi}$ is shown in figure 8 together with the number actually observed.

95% confidence level limits of

$$-1.9 \le \alpha_{W\phi} \le +2.0,$$

$$-1.1 \le \tilde{\alpha}_W \le +1.3,$$

were obtained using a likelihood fit to the product of Poisson probabilities of seeing the observed number of events in each channel given the expected number as a function of the TGC. As the different channels can have additional TGC vertices, a four-fermion analysis was used throughout and the channels were not constrained to the SM branching ratios. Systematic contributions from the uncertainty in M_W, the uncertainty in the LEP energy, the background cross-sections and Monte Carlo statistical errors on the efficiencies were evaluated and are included in the above limits.

Analysis of the 172 GeV data is in progress.

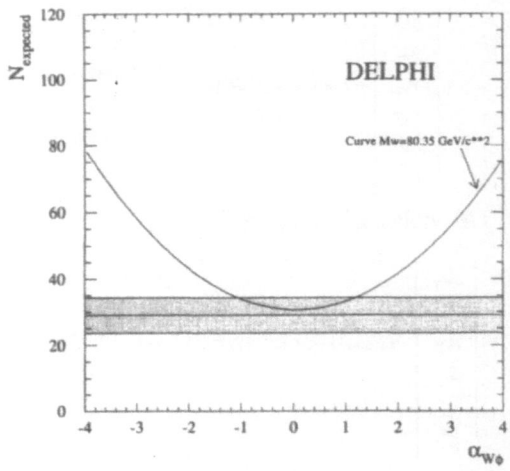

Figure 8. Expected number of events as a function of $\alpha_{W\phi}$. The curve is the prediction for $M_W = 80.35$ GeV/c²; the horizontal line is the observed number of events and the shaded band its one standard deviation errors.

CONCLUSIONS

DELPHI has measured the WW cross-section at 161.3 GeV to be

$$\sigma_{WW}^{total} = 3.67^{+0.97}_{-0.85} \text{ (stat.)} \pm 0.19 \text{ (syst.) pb,}$$

corresponding to a W mass of

$$M_W = 80.40 \pm 0.44 \text{ (stat.)} \pm 0.09 \text{ (syst.)} \pm 0.03 \text{ (LEP) GeV/c}^2.$$

The cross-section measurement was also interpreted in terms of limits at 95% confidence level on non-standard trilinear gauge couplings

$$-1.9 \leq \alpha_{W\phi} \leq +2.0,$$
$$-1.1 \leq \bar{\alpha}_W \leq +1.3.$$

A preliminary combination of the results of the four LEP collaborations gave

$$\sigma_{WW}^{total} = 3.65 \pm 0.45 \text{ pb (LEP preliminary),}$$

corresponding to

$$M_W = 80.42^{+0.22}_{-0.21} \text{ GeV/c}^2 \text{ (LEP preliminary),}$$

in good agreement with measurements from $p\bar{p}$ experiments and from fits to LEP1 data.

DELPHI has also measured the WW cross-section at 172 GeV to be

$$\sigma_{WW}^{total} = 11.38^{+1.54}_{-1.43} \text{ (stat.)} \pm 0.32 \text{ (syst.) pb (DELPHI preliminary),}$$

which is in good agreement with Standard Model predictions.

Analysis of the data taken at 172 GeV is proceeding and results from the direct reconstruction of M_W and limits on TGC's from differential cross-sections are expected soon. In 1997, LEP is scheduled to run at a higher centre-of-mass energy, circa 184 GeV, and we hope to accumulate at least 70 pb^{-1} per experiment, allowing an estimated precision of $\Delta M_W \sim 70$ MeV and improved limits on non-standard TGC's.

References

[1] M. Rijssenbeek, *W mass from the Tevatron*, FERMILAB CONF-96/365-E, to appear in the proceedings of the 28th International Conference on High Energy Physics, Warsaw, 25-31 July 1996.

[2] Z. Kunszt and W. J. Stirling (Convenors), *Determination of the mass of the W boson*, in: *Physics At LEP2*, G. Altarelli, T. Sjöstrand and F. Zwirner (eds.), CERN 96-01 (1996).

[3] F. Abe et al. (CDF collaboration), *Phys. Rev. Lett.* 74(1995) 1936.

[4] F. Abe et al. (CDF collaboration), *Phys. Rev. Lett.* 75(1995) 1017.

[5] S. Abachi et al. (D0 collaboration), *Phys. Rev. Lett.* 77(1996) 3033.

[6] S. Abachi et al. (D0 collaboration), *Limits on anomalous $WW\gamma$ couplings from $p\bar{p} \rightarrow W\gamma + X$ events at $\sqrt{s} = 1.8$ TeV* Fermilab-Pub-96/434-E, hep-ex/96120002 (1996).

[7] G. Gounaris, J.-L. Kneur, D. Zeppenfeld (Convenors), Triple Gauge Boson Couplings, in: *Physics At LEP2*, G. Altarelli, T. Sjöstrand and F. Zwirner (eds.), CERN 96-01 (1996).

[8] P. Abreu et al. (DELPHI collaboration), Measurement and interpretation of the W-pair cross-section in e^+e^- interactions at 161 GeV, CERN-PPE97/09, to appear in Physics Letters B.

[9] C. G. Papadopoulos, *ERATO: event generator for four-fermion production at LEP2 energies and beyond*, DEMO-HEP-96/02, hep-ph/9609329 (1996) to appear in *Comp. Phys. Comm.*.

[10] F. A. Berends, R. Kleiss, R. Pittau, EXCALIBUR in: *Physics At LEP2*, G. Altarelli, T. Sjöstrand and F. Zwirner (eds.), CERN 96-01 (1996).

[11] D. Bardin et al., GENTLE/4fan in: *Physics At LEP2*, G. Altarelli, T. Sjöstrand and F. Zwirner (eds.), CERN 96-01 (1996).

[12] LEP Electroweak Working Group, private communication.

[13] V. C. Spanos and W. J. Stirling, *Constraining a CP-violating WWV coupling from the W^+W^- threshold cross section at LEP2*, Durham preprint DTP/96/54, hep-ph/9607420 (1996).

INDEX

SUGRA, 93, 95
SU(N), 97, 99, 187
SUSY, 81, 85, 95, 133
SUSY breaking, 85, 86
SUSY WIMPS, 81, 85, 88, 134, 144

T

Tau netrinos, 22
Time reversal, 67
Two superbody problems, 174

W

W mass, 230

W physics, 224
Weyl representation, 158
Witten, 166
Wolfenstein parametrization, 74

X

^{139}Xe
Xenon detector, 141

Y

Yang–Mills, 182
Yukawn couplings, 69, 95
Yukawa theory, 196